BC1

AF689684

Elementary Mathematics of Price Theory

Clark Lee Allen

Southern Illinois University

Elementary Mathematics of Price Theory

WADSWORTH PUBLISHING COMPANY, INC.

Belmont, California

© 1962 by Wadsworth Publishing Company, Inc., Belmont, California. All rights reserved. No part of this book may be reproduced in any form, by mimeograph or any other means, without permission in writing from the publisher.

L.C. Cat. Card No.: 62-9812

Printed in the United States of America

Sixth printing: August 1967

To Bernice

Contents

Foreword

This book is based on the premise that one of the basic steps in the study of price theory is the mastery of the tools used by economists who write in the area of micro economics. These tools are essentially mathematical in nature, even though some writers attempt to disguise their mathematical character by stating concepts verbally instead of in conventional symbols.

The mathematical backgrounds of those who study price theory vary greatly. Because of this diversity, some instructors avoid the use of mathematical tools almost entirely, and others proceed to use the most effective tools for the purpose at hand, apparently expecting students to follow the analysis on the basis of some sort of intellectual osmosis. Even those students whose records would seem to indicate an adequate background in mathematics often appear to have difficulty in applying what they presumably have learned to the problems of economics. The result is that after having completed instruction in price theory they still do not have an analytical grasp of the area, and most of them are poorly equipped to read the literature with any understanding.

One of the advantages of academic vagabondage is that it increases the scope of pedagogical experimentation. The author has taught price theory at Duke University, Northwestern University, Florida State University, Texas A & M College, North Carolina State College, and Southern Illinois University. The materials in this book were used in essentially the present form in the last three institutions. The mathematical backgrounds of the students who have used these materials have ranged from no high school mathematics, except business arithmetic, to a graduate major in mathematics with an economics minor. The course appeared to be as valuable to one as to the other.

This book is meant to provide the basic mathematics needed by the nonmathematical economist. It is designed for use in a course in economics rather than a course in mathematics. A number of students who have used these materials have subsequently elected courses in mathematical economics, and it seems clear that some of these students would not have undertaken that work if they had not first studied these materials. But the book is designed for those students who need more work with the basic tools of economic analysis, and that includes most of them regardless of their interests in mathematical economics and econometrics.

Although this book has deliberately been kept short in order to make it manageable within the limits of a single course, a considerable variety of topics has been included, and it is believed that the techniques of prime importance to the price theorist are dealt with here. There is, however, no clearly logical stopping place in matters of this kind, and some instructors may find that subjects of particular interest to them have been left out. Although it is easier to err on the side of attempting too much rather than too little in courses of this sort, if the problem does arise, it can be handled in two ways: supplementary reading in other books can be assigned, or the instructor may try his hand at preparing additional chapters for the use of his classes. Among the topics that might be treated in this manner are logarithmic functions and their derivatives, determinants and matrix notation, difference equations, and the mathematical applications to problems in aggregate economic analysis. The basic groundwork for these topics is included here, but the book has been limited to those techniques that seem to be most useful to the student of price theory.

A second premise of this book is that the student needs to know something of both calculus and geometry. If it is assumed that students know no calculus, textbook writers have no choice but to limit their analyses to the geometry of price theory. Those who make use of the calculus, however, appear to assume that if the student knows calculus, there is little need for geometric proofs. The experience of the author suggests that the use of geometry provides insights that are otherwise lacking even for those skilled in the use of the calculus. And since economists devote a considerable part of their time trying to explain economic matters to those who know no calculus, it would appear that some proficiency in the geometry of price theory is important for that purpose alone.

Parts *A* and *B* lay no claim to originality. Materials are presented in orthodox fashion, but a careful effort has been made to move in short, easy steps in order to instill confidence in the student to work with mathematical symbols. Part *C* contains some demonstrations that have not previously appeared in the literature, but the primary object here has been to show the student that much can be done with simple geometry, even though the work is often more cumbrous than when the more powerful mathematical tools are employed. The Appendix includes a review of mathematics fundamentals that in the body of the text are assumed to be familiar to the student.

To the many students who willingly subjected themselves to experiments, some of which didn't work out, who studied from imperfect mimeographed materials without a murmur, who often stubbornly refused to perceive the obvious and then grasped the subtle without a struggle, the author expresses a thousand thanks.

C. L. A.

Bibliography

The books referred to in the bibliographical note at the end of each chapter are listed here; journal articles cited are not included.

Allen, Clark Lee, James M. Buchanan, and Marshall R. Colberg, *Prices, Income, and Public Policy*, 2nd ed. (New York: McGraw-Hill, 1959).

Allen, Clark Lee, Aurelius Morgner, and Robert H. Strotz, *Problems in the Theory of Price* (Englewood Cliffs, N.J.: Prentice-Hall, 1954).

Allen, R. G. D., *Mathematical Analysis for Economists* (London: Macmillan, 1942).

Allen, R. G. D., *Mathematical Economics* (London: Macmillan, 1956).

Bach, George Leland, *Economics, an Introduction to Analysis and Policy*, 3rd ed. (Englewood Cliffs, N.J.: Prentice-Hall, 1960).

Bain, Joe S., *Price Theory* (New York: Holt, 1952).

Bober, M. M., *Intermediate Price and Income Theory* (New York: Norton, 1955).

Boulding, Kenneth E., *Economic Analysis*, 3rd ed. (New York: Harper, 1955).

Boulding, Kenneth E., and W. Allen Spivey, *Linear Programming and the Theory of the Firm* (New York: Macmillan, 1960).

Brown, E. H. Phelps, *The Framework of the Pricing System* (Lawrence, Kan.: Student Union Book Store, 1949).

Bushaw, D. W., and R. W. Clower, *Introduction to Mathematical Economics* (Homewood, Ill.: Irwin, 1957).

Cassel, Gustav, *The Theory of Social Economy*, rev. ed. (New York: Harcourt Brace, 1931).

Chamberlin, Edward Hastings, *The Theory of Monopolistic Competition*, 7th ed. (Cambridge: Harvard, 1957).

Clemence, Richard V., *Readings in Economic Analysis*, Vol. II (Reading, Mass.: Addison-Wesley, 1950).

Crum, W. L., and Joseph A. Schumpeter, *Rudimentary Mathematics for Economists and Statisticians* (New York: McGraw-Hill, 1946).

Daus, Paul H., and William M. Whyburn, *Introduction to Mathematical Analysis* (Reading, Mass.: Addison-Wesley, 1958).

Davidson, Ralph K., Vernon L. Smith, and Jay W. Wiley, *Economics: an Analytical Approach* (Homewood, Ill.: Irwin, 1958).

Dorfman, Robert, Paul A. Samuelson, and Robert M. Solow, *Linear Programming and Economic Analysis* (New York: McGraw-Hill, 1957).

Due, John F., *Intermediate Economic Analysis*, 3rd ed. (Homewood, Ill.: Irwin, 1956).

Enke, Stephen, *Intermediate Economic Theory* (Englewood Cliffs, N.J.: Prentice-Hall, 1950).

Fellner, William, and Bernard F. Haley, ed., *Readings in the Theory of Income Distribution* (Philadelphia: Blakiston, 1946).

Fisher, Irving, *A Brief Introduction to the Infinitesimal Calculus* (New York: Macmillan, 1937).

Hicks, J. R., *Value and Capital*, 2nd ed. (Fair Lawn, N.J.: Oxford U.P., 1946).

Knight, Frank H., *Risk, Uncertainty, and Profit* (Boston: Houghton Mifflin, 1921).

Leftwich, Richard H., *The Price System and Resource Allocation*, rev. ed. (New York: Holt, Rinehart and Winston, 1960).

Lerner, Abba P., *The Economics of Control* (New York: Macmillan, 1944).

Makower, Helen, *Activity Analysis and the Theory of Economic Equilibrium* (London: Macmillan, 1956).

Marshall, Alfred, *Principles of Economics*, 8th ed. (London: Macmillan, 1938).

McKenna, Joseph P., *Intermediate Economic Theory* (New York: Dryden, 1958).

Norris, Ruby Turner, *The Theory of Consumer's Demand* (New Haven: Yale, 1947).

Richardson, M., *Fundamentals of Mathematics*, rev. ed. (New York: Macmillan, 1958).

Robinson, Joan, *Collected Economic Papers* (New York: Kelley, 1951).

Robinson, Joan, *The Economics of Imperfect Competition* (London: Macmillan, 1942).

Schultz, Henry, *The Theory and Measurement of Demand*, 2nd ed. (Chicago: U. of Chicago, 1957).

Scitovsky, Tibor, *Welfare and Competition* (Homewood, Ill.: Irwin, 1951).

Shackle, G. L. S., *Economics for Pleasure* (New York: Cambridge U.P., 1959).

Stigler, George J., *The Theory of Price*, rev. ed. (New York: Macmillan, 1954).

Stigler, George J., and Kenneth E. Boulding, ed., *Readings in Price Theory* (Homewood, Ill.: Irwin, 1952).

Stonier, Alfred W., and Douglas C. Hague, *A Textbook of Economic Theory* (New York: Longmans, Green, 1954).

Tintner, Gerhard, *Mathematics and Statistics for Economists* (New York: Rinehart, 1953).

Walras, Léon, *Elements d'économie politique pure*, 4th ed. (Lausanne, 1900). Translated by William Jaffé, *Elements of Pure Economics* (Homewood, Ill.: Irwin, 1954).

Weintraub, Sidney, *Price Theory* (New York: Pitman, 1956).

A

Functions and Graphs

1

Functions and Their Diagrammatic Representation

1.1 The Use of Symbols

Man's cultural and scientific advances have been the result, in very large part, of his ability to devise and use symbols to assist him in his thinking processes. Language is made up of words, and words are symbolic representations of mental images. Primitive man undoubtedly used elemental thought processes before the invention of language, but his intellectual achievements were of necessity narrowly circumscribed before he was able to attach certain sounds to given ideas. The invention of written language greatly facilitated man, not only in communicating ideas, but also in developing them. The use of numbers, which, of course, are also symbols, greatly widened man's mental horizons. In virtually all of the contemporary arts and sciences, symbols of a great variety are employed. Some of these symbols take the form of words, and they are referred to as concepts or technical terms. The girl who goes to a baseball game and suffers from misconceptions concerning such expressions as "hit a foul," "steal a base," and "slide home" is legendary. Symbols frequently assume numerical, algebraic, or geometric forms. Or special symbols suitable for the purpose may have to be invented; musical notation that makes it possible for two symphony orchestras to produce essentially identical musical effects from the same score, and diagrams of electrical circuits used by physicists and electrical engineers, are cases in point. The more complex the mental process involved, the greater reliance we have to place on symbols of one sort or another.

Imagine trying to design an interplanetary missile without an elaborate system of symbols!

In economic analysis a variety of symbols is employed. Verbal concepts such as labor, profits, income, distribution, prices, and costs are essential. For the most part, economics—unlike most of the biological and physical sciences—uses everyday, homely words in its scientific vocabulary; but the beginning student of economics must early learn the special meanings that economists attach to familiar concepts. Land, for example, does not mean in economics precisely what it means in ordinary speech. But in addition to the verbal concepts employed by economists, many geometric charts are employed to understand and to explain economic phenomena. Demand and supply curves and cost curves, for example, are familiar to all beginning students of economics, and their use is justified on the grounds that they help us to think straight, and provide us with insights that would be difficult to perceive without their help.

Economic analysis in general and price theory in particular are concerned with quantities and quantitative relationships. Price, quantity demanded, total revenue, and average cost—all of these are quantities. Marginal revenue and marginal cost represent rates of change in quantitative magnitudes. Elasticity of demand is a complex relationship between changes in quantity demanded and changes in price. Since the economist, by the nature of the subject matter he studies, is concerned with relationships between quantitative magnitudes, much of his science is mathematical in nature. Stating propositions verbally without the use of conventional mathematical symbols does not change the essentially mathematical character of the discipline. Marginal cost, for example, is a mathematical concept whether stated verbally or in mathematical symbols. The idea of profit maximization is also a mathematical concept. Much of economic analysis is inescapably mathematical in nature.

Because of the essentially quantitative nature of much of economics, many economists have been able to apply powerful mathematical tools to economic data, and particularly in recent years the school of mathematical economists and econometricians has been growing in influence and numbers. These writers have not been without their critics. It has been charged that mathematical economists are interested in only the quantitative aspects of the subject, that they become so interested in mathematical manipulations that they lose sight of the essentially economic aspects of the subject, and that they belittle researches of a nonmathematical sort. But regardless of the merits of the issues between the mathematical and the nonmathematical economists, there remains a certain minimum competence in mathematics that is required of all economists, the nonmathematical (or even the antimathematical) economists included.

One final introductory observation. Symbols are man-made and quite arbitrary. To a considerable degree advancements in science depend upon the invention of ways to measure things. Prior to the invention of the thermometer one might have observed that today is colder than yesterday, but there would have been no way to indicate how much colder. But note how arbitrary and devoid of glamor a thermometer is. Let us call the freezing point of water 0° and the boiling point of water 100°, and we have the basis of the centigrade thermometer. In the centuries before the invention of the thermometer it would have made no more sense to ask how much hotter A is than B than it is currently to make interpersonal comparisons of utility. Other units of measurement are equally arbitrary. How long is a yard? To say that it is three feet or 36 inches is no help if we don't know how long an inch or a foot is. A yard is the length of the standard yard, which is a piece of metal compensated for even the slightest changes in temperature; that is to say, a yard is the length of a piece of metal that has arbitrarily been designated as a yard. Similarly, a dollar is currently defined as $\frac{1}{35}$ of an ounce of gold.

Even our system of counting is quite arbitrary. It appears to be generally assumed that $2+2=4$ by virtue of some sort of natural law, but the fact is that this is true because we use a ten-scale or decimal system of numbers. Suppose that instead of zero plus nine digits we used zero plus only three digits. We would then count as follows:

$$\begin{gathered} 0,\ 1,\ 2,\ 3, \\ 10,\ 11,\ 12,\ 13, \\ 20,\ 21,\ 22,\ 23,\ \cdots. \end{gathered}$$

In this event,

$$\begin{aligned} 2+2 &= 10, \\ 13+1 &= 20, \\ 21 \div 3 &= 3. \end{aligned}$$

It has been suggested that we use the ten-scale system because we have two hands and a total of ten fingers. If we had only one hand and five fingers, it is possible that a five-scale system might have been adopted. With such a scale, as the student may confirm for himself, the following would be valid:

$$1+4 = 10,$$
$$2+3 = 10,$$
$$3+4 = 12,$$
$$4+2 = 11.$$

Actually one system is about as good as another except that we are familiar with the decimal scale. For some purposes in higher mathematics, other systems have certain advantages. Some electronic computers, for example, use the binary digit (abbreviated *bit*) system; only the two digits *0* and *1* are employed.

1.2 Functions

We turn now to a matter of somewhat more immediate concern to the task at hand. In ordinary speech we frequently hear it said that one thing is a function of another. *Function* as used in this sense does not mean an elaborate and impressive ceremony, like the senior prom, or the normal action of a part of a plant or animal, like the function of the liver, or a profession or occupation, like the clerical function. If we say that one variable is the function of another variable, we mean that one magnitude is related to another magnitude in such a way that for each value of the latter there corresponds a unique value of the former. When we say that the quantity demanded is a function of price, we mean that for every price there is a corresponding quantity that will be demanded. This may be put somewhat more formally as follows: the variable y is said to be a function of the variable x when y depends on x in such a way that the fixing of x determines a corresponding value of y. If we want to indicate that y is a function of x without fixing the specific form of the function, we write $y=f(x)$, which is read, "y is a function of x." This simply means that for every value of x there is a corresponding value of y. Instead of f we may use as function symbols such letters as g, h, F, G, and so on. Or we may write $y=y(x)$.

Suppose we have the function $y=f(x)=1/(2x)$.

By making the appropriate substitution for x we can determine the corresponding value of y. If $x=3$, then $y=1/(2\cdot 3)=\frac{1}{6}$. If $x=-5$, $y=-1/10$. If $x=50$, $y=1/100$. For all permissible values of x (that is, all values of x except $x=0$—dividing by zero is not a permissible operation) there is a corresponding value for y; y is a function of x.*

When we substitute specific values for x in the function $y=f(x)=1/(2x)$, we may write $f(3)=\frac{1}{6}$, which means that x has been replaced by 3 in the function; $f(-3)=-\frac{1}{6}$; $f(\frac{3}{2})=1/(\frac{6}{2})=\frac{1}{3}$; $f(1/a)=1/2(1/a)=1/(2/a)=a/2$.

1.3 Exercises

1. Find $f(0), f(1), f(-1), f(-2)$, if

$$f(x) = x^3-3x^2+5x-1.$$

2. Find $F(0)$, $F(\frac{1}{2})$, $F(-\frac{1}{2})$, if

$$F(x) = x(1-4x^2).$$

3. Let $f(x)=3x$. Find $f(a)$; $f(b)$; $f(a+b)$. Show that in this instance

$$f(a+b) = f(a)+f(b).$$

4. Let $f(x)=ax^2+bx+c$, where a, b, and c are constants. Find $f(0)$; $f(-1)$; $f(b-a)$. Show that

$$f(b-a) \neq f(b)-f(a).$$

1.4 Graphs of Equations

1. Let $y=f(x)=4x-3$. Fill in the values of y in the table below:

x	$y = f(x) = 4x-3$
-4	
-3	
-2	
-1	
0	
1	
2	
3	
4	

* The reason for excluding dividing by zero as a permissible operation can be indicated by the following demonstration:

Let

$a = b$	(1)	
$a^2 = ab$	(2)	multiplying both sides by a
$a^2-b^2 = ab-b^2$	(3)	subtracting b^2 from both sides
$(a+b)(a-b) = b(a-b)$	(4)	factoring
$a+b = b$	(5)	dividing both sides by $(a-b)$
$b+b = b$	(6)	since $a = b$
$2b = b$	(7)	
$2 = 1$	(8)	dividing both sides by b.

The error in this analysis comes in step (5), where we divided both sides of the equation by $(a-b)$. Since $a=b$ (step 1), dividing by $(a-b)$ is dividing by zero.

2. Let $y=f(x)=2x^2+8$. Fill in the values of y in the table below:

x	$y=f(x) = 2x^2+8$
-4	
-3	
-2	
-1	
0	
1	
2	
3	
4	

3. On separate charts plot the functions $y=4x-3$ and $y=2x^2+8$.

4. Plot on the same coordinate system the functions $y=x^2$ and $y=6-x$ for the range from $x=-4$ to $x=4$. Determine graphically the points of intersection of the curves. Confirm algebraically.

1.5 Bibliographical Note*

A discussion of functions and graphs will be found in almost any introduction to mathematics textbook. M. Richardson, *Fundamentals of Mathematics*, rev. ed., pp. 275–284, is a good one. Gerhard Tintner, *Mathematics and Statistics for Economists*, Chapter 1, is excellent. The technique for constructing a graph from an equation is spelled out in some detail in Clark Lee Allen, Aurelius Morgner, and Robert H. Strotz, *Problems in the Theory of Price*, pp. 141–150, and Clark Lee Allen, James M. Buchanan, and Marshall R. Colberg, *Prices, Income, and Public Policy*, 2nd ed., pp. 467–473. An excellent discussion of functions and their diagrammatic representation may be found in R. G. D. Allen, *Mathematical Analysis for Economists*, pp. 28–50. See also D. W. Bushaw and R. W. Clower, *Introduction to Mathematical Economics*, Chapter 8.

A symposium on mathematical economics, which presents a variety of views on the use of mathematics in economics, appears in the *Review of Economics and Statistics*, 1947, pp. 269 ff, and this can be read by the student with profit. The difficulties of presenting essentially mathematical concepts without the use of mathematical symbols are illustrated in G. L. S. Shackle, *Economics for Pleasure*, especially Chapters 8 and 9.

A brief discussion of nondecimal-scales of notation will be found in Richardson, *op cit.*, pp. 143–147.

* A bibliographical note concludes each chapter. For the complete bibliographical data for the books cited, consult the Bibliography, which follows the Table of Contents.

2

Linear Equations

2.1 Linear Equations

Economists frequently find it convenient to represent relationships graphically. The demand for a commodity, for example, may be represented by a demand "curve." Since straight lines are somewhat easier to manipulate than nonlinear curves, demand is frequently represented as being linear. If we let x be the quantity demanded and y be price, we know that $x=f(y)$. The specific demand for a particular commodity may be given; for example, we may assume that

$$x = f(y) = 10-y.$$

It is an easy matter to represent this function graphically. If $y=0$, $x=10$; and if $x=0$, $y=10$. This means that, when plotted, the demand curve will cut the x-axis at 10 and the y-axis at 10. This is the demand curve corresponding to the function $x=f(y)=10-y$.

By the use of some of the rudiments of analytical geometry it is possible to move in the opposite direction. Suppose we have given a linear demand curve and we want to determine the equation for the curve. The devices described in this chapter make it possible to do this.

But first a word about equations. An equation is a statement of equality between two quantities. The equation may be an identity; for example, the equation

$$5(a+b) = 5a+5b$$

is an identity since it is true for all values of a and b. Consider, however, the equation

$$3x-1 = 8.$$

This equation is not an identity because it is true only for the value $x=3$. This equation contains only the first power of the unknown, and, when plotted, it is represented by a straight line. For this reason it is called a linear function. The general form for a linear function is

$$y = ax+b,$$

where a and b are constants.

2.2 Exercises

1. Solve for x: $y=x+3$.
2. Solve for x: $2x-5y+8=0$.
3. Solve for x: $6(x+1)+2(-x+2)-(x-2)=0$.
4. Remembering that the first number in the parentheses is always the x-value of the point and the second number is the y-value, determine whether the points (1, 1), (5, −5), (0, $\frac{5}{3}$), (13, 6) lie on the curve $2x+3y=5$.
5. Plot the function $x+2y-4=0$.
6. Plot the function $y=5$.
7. Find the point of intersection of the curves $3x-y=1$ and $x+2y=5$.
8. Plot the function $12x-6xy=0$.

2.3 A Line through Two Given Points

Consider a line to be drawn through two given points. The two points determine the particular line without ambiguity. The equation for a line that passes through the points (x_1, y_1) and (x_2, y_2) is given by

$$y-y_1 = \left(\frac{y_2-y_1}{x_2-x_1}\right)(x-x_1).$$

Consider, for example, the straight line that goes through the points (1, 1) and (3, 7). We have $x_1=1$; $y_1=1$; $x_2=3$; and $y_2=7$. Substituting in the above equation, we get

$$y-1 = \left(\frac{7-1}{3-1}\right)(x-1).$$

Simplifying this we have

$$y = 3x-2.$$

2.4 Exercises

Write the equations of the lines determined by the following pairs of points:

1. (3, 2), (−3, 1).

2. (0, 3), (5, 0).

3. (0, 0), (−1, −4).

2.5 Line through a Given Point with a Given Slope

The expression

$$m = \frac{\Delta y}{\Delta x} = \frac{y_2 - y_1}{x_2 - x_1}$$

is known as the slope or gradient of the line. For example, as we move from one point on a straight line to another point on the same straight line, if we move 3 units in the y-direction for every unit we move in the x-direction, the slope of the line is 3. If the curve slopes upward to the right, the slope is positive; if the curve slopes downward to the right, its slope is negative. If we know one point on a straight-line curve and the slope of the curve, the line is determined without ambiguity. The equation

$$y - y_1 = m(x - x_1)$$

gives a straight line that passes through the point (x_1, y_1) with a slope of m.

Consider the line that passes through the point (−2, 4) with a slope of 2. The point-slope formula gives us

$$(y-4) = 2[x-(-2)] = 2(x+2)$$
$$y = 2x+8.$$

It can be readily verified that the coefficient a in the function $y = ax + b$ is the slope m. So we may write the equation of a straight line in the form

$$y = f(x) = mx + b.$$

By taking $x = 0$, we find that

$$f(0) = b.$$

Hence b is the value of y, which corresponds to $x = 0$; that is, b is the intercept of the curve on the y-axis. The equation $y = mx + b$ is called the slope-intercept equation for a straight line.

2.6 Exercises

1. Write the equations of the lines that pass through the given points with the indicated slopes:

   (a) $(0, -4)$, $m=\frac{1}{2}$.
   (b) $(-1, -5)$, $m=-\frac{1}{5}$.
   (c) $(3, 4)$, $m=0$.

2. Write the equations of the lines that satisfy the following conditions:

   (a) x-intercept of 2 and y-intercept of -4.
   (b) $m=-3$, y-intercept of 2.
   (c) Passing through the point $(-2, 3)$ parallel to the line $4x-y+3=0$.

3. Find the slopes, x-intercepts, and y-intercepts of the following lines:

   (a) $2x-3y+12=0$.
   (b) $3x+4y+2=0$.

4. Find the value of k so that the line $2x-3ky+8=0$ will pass through point $(0, 2)$.

2.7 Bibliographical Note

The material in this chapter is elementary analytical geometry, and the points made here will be supplemented in any good textbook in that field. Or the student may find the following helpful: R. G. D. Allen, *Mathematical Analysis for Economists*, pp. 61–69; M. Richardson, *Fundamentals of Mathematics*, rev. ed., pp. 224–231, 233–241, and 247–251; Gerhard Tintner, *Mathematics and Statistics for Economists*, pp. 11–16.

3

Linear Demand and Supply Functions

3.1 Linear Demand Functions

Probably no concepts are more fundamental to price theory than demand and supply. These can be represented in any of several ways. In most general terms we may say simply that the quantity demanded is a function of price, or $D=f(p)$. Or we may specify a particular demand by saying, for example, that $D=f(p)=10-p$. Or this demand may be represented by a demand schedule, which consists of a table of two columns; the first column indicates the various possible prices, and the second column indicates the quantity demanded at each indicated price. Or, finally, the demand schedule may be plotted graphically as a demand curve.

Assume a market demand $D=f(p)=100-5p$. (a) What is the quantity demanded if the price is \$3? We have $D=f(3)=100-5(3)=85$. (b) Assume the quantity demanded is 20 units. What is the corresponding price? We have $20=100-5p$, and by solving this equation, it is found that $p=\$16$. (c) What would be the quantity demanded if the commodity were a free good, *i.e.*, if $p=0$? We have $D=100-5(0)=100$. (d) What is the highest price anyone will pay for this commodity? Put $D=0=100-5p$. Then $p=\$20$. Actually $D=f(20)=0$; the price must be something less than \$20 if any of the commodity is to be sold on the market.

3.2 Exercises

1. $D = 25 - 5p$ is a market demand curve. Find the price if the quantity demanded is

(a) 20;
(b) 15;
(c) 5 units.

Find the quantity demanded if the price is

(d) 3;
(e) 2;
(f) 1.
(g) Find the quantity demanded if the commodity is a free good.
(h) What is the highest price anyone would pay for this commodity?
(i) Plot the demand curve.

2. The demand curve for sugar in the U.S. for 1915–1929 was estimated by Henry Schultz to be $D = 135 - 8p$. Find the quantity demanded if the price is

(a) 3;
(b) 10;
(c) 5.

What is the price corresponding to a demand of

(d) 80?
(e) 95?
(f) 35?
(g) What is the highest price anyone would pay for sugar?
(h) How much sugar would be consumed if it were a free good?
(i) Plot the demand curve.

3. The demand for a commodity is $D = a - bp$, where a and b are positive constants.

(a) Find the price if the quantity demanded is $a/2$.
(b) Find the quantity demanded if the price is $a/3b$.
(c) Find the amount demanded if the commodity is a free good.
(d) What is the highest price anyone will pay for it?

3.3 Linear Supply Functions

Let $S = f(p) = 6p - 2$ be the supply curve for a commodity. (a) Given $p = \$5$, what is the quantity supplied? We have $S = f(5) = 6(5) - 2 = 28$. (b) What price will cause a supply of 16? The equation $16 = 6p - 2$ has the solution $p = 3$. (c) What is the lowest price at which the commodity will be supplied? We put $S = 0 = 6p - 2$, which gives $p = \frac{1}{3}$. Actually,

$S=f(\frac{1}{3})=0$. At a price of $\frac{1}{3}$ nothing will be supplied; hence the price must be somewhat greater than $\frac{1}{3}$ if any of the commodity is to be supplied.

3.4 Exercises

1. The supply of a commodity is $S=10p-40$. Find the price if the quantity supplied is

(a) 10;
(b) 20;
(c) 100.

Find the quantity supplied if the price is

(d) 5;
(e) 7.50;
(f) 50.
(g) What is the lowest price at which any of the commodity will be supplied?
(h) Plot the supply curve.

2. The supply of a commodity is $S=Ap-B$, where A and B are positive constants. Find the price if the quantity supplied is

(a) $5A-B$;
(b) $A+2B$.

Determine the quantity supplied if the price is

(c) $3B/A$;
(d) $B/5A$.
(e) What is the lowest price that will cause any supply of the commodity?

3. Henry Schultz estimates the supply curve for imported sugar for the U.S. in 1903–1913 as $S=1.1p-0.1$. Find the price if the quantity supplied is

(a) 1;
(b) 0.8;
(c) 0.5.

Determine the quantity supplied if the price is

(d) 8;
(e) 6;
(f) 4.1.
(g) What is the lowest price that will cause any sugar to be supplied?
(h) Plot the supply curve.

3.5 Equilibrium Price

Suppose that the demand for a commodity is $D=10-2p$ and the supply is $S=5p-4$. Equilibrium price is established where demand equals supply, *i.e.*, where $D=S$ or where $10-2p=5p-4$. Solving this equation we get for the equilibrium price $p=2$. Inserting this value of p in the demand equation we have $D=10-4=6$. Substituting the same value in the supply equation we have $S=10-4=6$. The equilibrium price, therefore, is 2 and the quantity exchanged is 6.

3.6 Exercises

1. Let the demand for a commodity be $D=16-4p$ and the supply be $S=6p-20$.
   (a) Find the equilibrium price.
   (b) Find the quantity exchanged.

2. Let the demand curve for a commodity be $D=a-bp$ and the supply curve be $S=mp-n$, where a, b, m, and n are positive constants.
   (a) Find the equilibrium price.
   (b) Determine the quantity exchanged.

3. Tintner estimated the demand for agricultural products in the U.S. from 1920–1943 to be $D=224.125-0.097p$, and the supply to be $S=-49.375+1.721p$.
   (a) Find the equilibrium price and quantity.
   (b) Plot the supply and demand functions.

3.7 Effect of Taxes and Subsidies

Let the demand curve of a commodity be $D=100-p$, and let the supply curve be $S=p-10$. By setting $D=S$, we get the following equilibrium values: $p=55$ and $D=S=45$. Assume now that a tax of \$20 per unit is imposed on the seller. The new supply curve becomes $S=(p-20)-10=p-30$. Again putting $D=S$, we find that $p=65$ and $D=S=35$. The total amount of the tax paid to the government is $20\times 35=700$. The price realized by the seller is $65-20=45$.

Now let us assume a subsidy of \$10 per unit. The new supply curve becomes $S=(p+10)-10=p$. When $D=S$, $p=50$, and the quantity exchanged is 50. The total amount of the subsidy is $50\times 10=500$. The price realized by the sellers is $50+10=60$.

3.8 Exercises

1. (a) Plot on the same axes $D=100-p$ and $S=p-10$, and indicate the equilibrium price and quantity. (b) Assume that a tax of \$20 per unit is imposed on the seller; draw the new supply curve and indicate the new equilibrium price and quantity. (c) Assume that instead of a tax a subsidy of \$10 is paid to the producers. Draw the appropriate supply curve and indicate equilibrium price and quantity.

2. Let the demand curve be $D=10-p$ and the supply curve be $S=p$. Determine the equilibrium quantity and price (a) before and (b) after the imposition on the seller of a specific tax of t per unit. (c) Suppose that the tax is levied on the buyer instead of on the seller, and the demand curve after the tax becomes $D=10-(p+t)$. Compare the effects of the tax on price and quantity when it is levied on the buyer and when it is levied on the seller. (d) Demonstrate graphically.

3.9 Bibliographical Note

The student at this point would do well to refresh his knowledge of demand, supply, and market price. He will find still fresh the classic treatment in Alfred Marshall, *Principles of Economics*, 8th ed., Book V, Chapters 1–3. Discussions of demand and supply in current elementary and intermediate economics textbooks (omitting for the moment discussions of elasticity) should be fruitful; those suggested for this purpose are Clark Lee Allen, James M. Buchanan, and Marshall R. Colberg, *Prices, Income, and Public Policy*, 2nd ed., Chapters 2–4; Richard H. Leftwich, *The Price System and Resource Allocation*, rev. ed., pp. 23–34; and Alfred W. Stonier and Douglas C. Hague, *A Textbook of Economic Theory*, Chapter 1. George J. Stigler, *The Theory of Price*, rev. ed., Chapter 2, will be particularly useful in connection with the problem on taxes and subsidies. A simple mathematical presentation of demand and supply will be found in Gerhard Tintner, *Mathematics and Statistics for Economists*, pp. 17–27.

4

General Equations for Linear Demand and Supply Functions

4.1 Linear Demand Functions

By analyzing the general form of linear demand functions we can discover certain relationships between the coefficients of x and y which make it possible to determine by inspection the x- and y-intercepts of the curves.

Let a = the x-intercept,
b = the y-intercept,

and A, B, and C = positive constants.

The general equation for a negatively inclined linear demand function may be written as:

$$Ax+By-C = 0.$$

Solving for y:

$$By = C-Ax$$

$$y = \frac{C}{B} - \frac{Ax}{B}. \qquad (1^a)$$

At $x=0$:

$$y = \frac{C}{B} = b.$$

Equation (1^a) may then be written:

$$y = b - \frac{Ax}{B}. \qquad (1^b)$$

At $y=0$:

$$Ax - C = 0$$

$$x = \frac{C}{A} = a.$$

The slope of the curve is equal to the y-intercept divided by the x-intercept. But since the slope of a demand curve is negative and both a and b are positive,

$$m = -\frac{b}{a} = \frac{-C}{B} \div \frac{C}{A} = \frac{-C}{B} \cdot \frac{A}{C} = \frac{-A}{B}.$$

Equation (1^b) becomes:

$$y = b + mx. \qquad (1^c)$$

To summarize: when the demand function is given in the general form,

$$a = \frac{C}{A},$$

$$b = \frac{C}{B},$$

$$m = \frac{-A}{B}.$$

4.2 Exercises

1. Determine a, b, and m for each of the following functions:

(a) $x+y-10=0$;
(b) $3x+2y-50=0$;
(c) $x=100-3y$;
(d) $Q=A(B-P)$ where Q=quantity and P=price;
(e) $x=(18-y)/2$.

2. (a) What is the general equation for a horizontal demand curve?
(b) What is the general equation for a vertical demand curve?

4.3 Linear Supply Functions

The general equation for a positively inclined linear supply function with a positive y-intercept may be written as:

$$Ax-By+C = 0.$$

Solving for y:

$$-By = -C-Ax$$

$$y = \frac{C}{B}+\frac{Ax}{B}. \tag{2^a}$$

At $x=0$:

$$y = \frac{C}{B} = b.$$

Equation (2^a) becomes:

$$y = b+\frac{Ax}{B}. \tag{2^b}$$

At $y=0$:

$$Ax+C = 0$$

$$Ax = -C$$

$$x = \frac{-C}{A} = a.$$

The slope of the curve is equal to the y-intercept divided by the x-intercept. But since the slope of a supply curve is positive and a is negative and b is positive,

$$m = \frac{-b}{a} = \frac{-C}{B} \div \frac{-C}{A} = \frac{C}{B}\cdot\frac{A}{C} = \frac{A}{B}.$$

Equation (2^b) becomes:

$$y = b+mx. \tag{2^c}$$

To summarize: when the supply function is given in the general form,

$$a = \frac{-C}{A},$$

$$b = \frac{C}{B},$$

$$m = \frac{A}{B}.$$

4.4 Exercises

1. Determine a, b, and m for each of the following supply functions:

(a) $x-y+10=0$;
(b) $3x-2y+60=0$;
(c) $Q=A(P-B)$ where $Q=$quantity and $P=$price;
(d) $x=100y-50$;
(e) $x=(y-18)/2$.

2. Prove that

$$\frac{x}{a}+\frac{y}{b} = 1$$

is the equation for a straight line with x-intercept$=a$ and y-intercept$=b$.

3. (a) What is the general equation for a linear supply curve with a positive x intercept?
(b) What is the value of a?
(c) of b?
(d) of m?

4. (a) Beginning with the equation for a straight line,

$$y = mx+b,$$

develop the general equation for a linear supply curve that passes through the origin.
(b) What is the value of a?
(c) of b?
(d) of m?

5. Demonstrate that (a) if a linear curve has one positive and one negative intercept, the slope of the line is positive, and (b) if both intercepts are either positive or negative, the slope of the line is negative.

6. Confirm the following alternate statement of the general expression for linear demand and supply functions: The general equation for a linear function is

$$Ax+By+C = 0.$$

(a) For a demand function, A and B are of the same sign and C is of opposite sign.
(b) For a supply function, A and B are of opposite sign and C is of arbitrary sign.

5

Elements of Nonlinear Functions

5.1 Components of a Curve

It is sometimes helpful to break up a nonlinear function of more than one term into its separate components and plot each term separately. Even a single number may be divided into its component parts. Consider the number 234. We may write:

$$234 = (2\cdot 10^2)+(3\cdot 10)+4.$$

If we let $x=10$, we may write:

$$234 = 2x^2+3x+4.$$

Since our system of counting is based on the ten-scale, any number may be represented in this way in terms of powers of 10.

Consider the function $y=\frac{1}{4}x^2+2x+15$. Plot on the same chart the following four curves over the range from $x=0$ to $x=6$:

1. $y=\frac{1}{4}x^2$.
2. $y=2x$.
3. $y=15$.
4. $y=\frac{1}{4}x^2+2x+15$.

5.2 Linear and Nonlinear Components

Next, we may separate the function into its linear and curved portions. Plot on the same chart the following curves from $x=0$ to $x=6$:

1. $y=\frac{1}{4}x^2$.
2. $y=2x+15$.
3. $y=\frac{1}{4}x^2+2x+15$.

5.3 Fixed and Variable Components

Finally, we may separate the function into its fixed and variable portions. Plot on the same chart the following curves from $x=0$ to $x=6$:

1. $y=\frac{1}{4}x^2+2x$.
2. $y=15$.
3. $y=\frac{1}{4}x^2+2x+15$.

These three curves may be taken to represent total variable cost, total fixed cost, and total cost.

5.4 Average Functions

If the total cost function is given, we may derive the average total cost by dividing by x. The three average cost functions corresponding to the total functions above would then take the form:

1. $AVC=\frac{1}{4}x+2$.
2. $AFC = \frac{15}{x}$.
3. $ATC = \frac{1}{4}x+2+\frac{15}{x}$.

Plot on the same chart these three curves from $x=0$ to $x=10$.

5.5 Bibliographical Note

For a discussion of the components of nonlinear functions see W. L. Crum and Joseph A. Schumpeter, *Rudimentary Mathematics for Economists and Statisticians*, Chapter 2.

This is an appropriate place to review the nature of cost functions. The following are appropriate for this purpose (omit the discussions of marginal cost for the moment): Clark Lee Allen, James M. Buchanan, and Marshall R. Colberg, *Prices, Income, and Public Policy*, 2nd ed., pp. 281–288; George Leland Bach, *Economics, an Introduction to Analysis and Policy*, 3rd ed., Chapter 20; Joe S. Bain, *Price Theory*, Chapter 3; John F. Due, *Intermediate Economic Analysis*, 3rd ed., Chapter 8; Richard H. Leftwich, *The Price System and Resource Allocation*, rev. ed., Chapter 8; George J. Stigler, *The Theory of Price*, rev. ed., Chapter 7.

B

The Calculus

6

Derivatives and Their Interpretation

6.1 The Concept of a Limit

We know that if $y=f(x)$, the value of y is determined by and changes with the value allotted to x. For many purposes it is important to be able to measure the *rate* at which y changes as x changes. This is the purpose of the differential calculus, which has been described as the "systematic consideration of the rates of increase of functions."

In developing the technique for the measurement of the rate of change of a function, the idea of a limit is important. Suppose that we have an unending sequence of numbers, $x_1, x_2, x_3, \cdots, x_n \cdots$. The term x_n is called the nth term or the general term. The sequence whose general term is $n/(n+1)$ is the sequence $\frac{1}{2}, \frac{2}{3}, \frac{3}{4}, \frac{4}{5}, \cdots, n/(n+1), \cdots$. The limit of the sequence is 1, because if we make n sufficiently large, we can get a value as close to 1 as we please. If, for example,

$$n = 1{,}000, \qquad \frac{n}{n+1} = \frac{1{,}000}{1{,}001};$$

or, if

$$n = 1{,}000{,}000, \qquad \frac{n}{n+1} = \frac{1{,}000{,}000}{1{,}000{,}001}.$$

The calculus has been described as being "concerned with the ultimate ratios of vanishing quantities." The meaning of this statement can be shown with the use of Figure 6-1. If we start at point P (OM, MP) and move to point S (ON, NS), we have moved PR units in the x-direction and RS units in the y-direction. We may draw a line from P through S. The slope of the line Pt is $\Delta y/\Delta x = RS/PR$. Now assume a smaller increment in x, MN'. The slope of the line Pt' which passes through the point S' on the curve is $R'S'/PR'$. If we next take an even smaller

Δx, MN'', the slope of the line Pt'', which passes through the point S'' on the curve, is $R''S''/PR''$. It will be noted that as Δx approaches zero, the line through P approaches TT', the tangent to the curve at point P. We may say that the tangent to the curve at P is the limit of $\Delta y/\Delta x$ as Δx approaches zero. This may be written

$$\text{tangent to the curve at } P = \lim_{\Delta x \to 0} \frac{\Delta y}{\Delta x}.$$

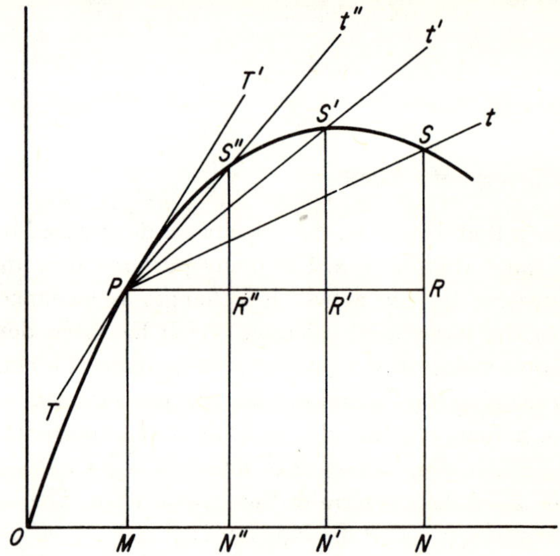

Fig. 6-1

6.2 The Derivative

Given the function

$$y = f(x),$$

continuous at the point P (x, y), let us assign to x an arbitrary increment Δx and compute the corresponding increment Δy of y. We have

$$y + \Delta y = f(x + \Delta x),$$

so that

$$\Delta y = f(x + \Delta x) - f(x).$$

Now form the ratio

$$\frac{\Delta y}{\Delta x} = \frac{f(x+\Delta x) - f(x)}{\Delta x}.$$

The limit of the ratio $\Delta y/\Delta x$ as Δx approaches zero is called the *derivative* of y with respect to x. The derivative is designated by the symbol dy/dx:

$$\frac{dy}{dx} = \lim_{\Delta x \to 0} \frac{\Delta y}{\Delta x} = \lim_{\Delta x \to 0} \frac{f(x+\Delta x)-f(x)}{\Delta x}.$$

Other symbols for the derivative are y' and $f'(x)$. The operation of finding the derivative is called *differentiation.*

The derivative of a function, as we will soon see, is identical with its rate of change. Geometrically, the derivative of a function is the slope of its graph. One of the important applications of the derivative in economic analysis is that the marginal function is the derivative of the total function. That is, if

$$\text{Total} = f(x),$$

then

$$\text{Marginal} = f'(x).$$

6.3 Example

The rate of speed at a given instant may be viewed as the average speed over an infinitesimally short period of time. If a motorist averages 50 miles per hour for one hour, it does not follow that he has traveled precisely at the *rate* of 50 miles per hour for the entire hour. His rate of speed may have been zero miles per hour while waiting for a traffic light, and out on the open road he may at times have gone as fast as 70 or 80 miles per hour. If we know that the driver has averaged 50 miles per hour for the last half hour, we still do not know at what rate he was traveling at any given instant of time. Even if we know the average speed for the last minute, we know little about the rate of speed; he may have accelerated during that minute from 30 to 90 miles per hour. But if we know the average speed for the last 1/10 second—or for the last 1/10,000 second—we may be sure that during that very short time interval the average speed and the rate of speed were virtually the same. What an accurate speedometer registers is the average speed for an infinitesimally short time period; this might be designated the marginal speed. During a very small time interval marginal and average speed are the same.

We know from physics that the distance traveled in a vacuum by a falling body is given by the equation

$$s = 16t^2,$$

where s is distance in feet and t is time in seconds. If, for example, a body has been falling for 2 seconds, it will have traveled $16 \times 4 = 64$ feet.

Let us now assume an arbitrary increase in t of Δt; the distance traveled by the falling body will then increase by some amount that we will designate as Δs. We may then set up the equation

$$s + \Delta s = 16(t + \Delta t)^2.$$

Solving,

$$s + \Delta s = 16(t^2 + 2t\Delta t + \Delta t^2)$$
$$s + \Delta s = 16t^2 + 32t\Delta t + 16\Delta t^2.$$

To determine the additional distance fallen in the additional time period, subtract s from $s + \Delta s$:

$$\Delta s = 32t\Delta t + 16\Delta t^2.$$

Dividing by Δt to get the average speed during the Δt time interval,

$$\frac{\Delta s}{\Delta t} = 32t + 16\Delta t.$$

Assuming that Δt approaches 0,

$$\frac{ds}{dt} = 32t.$$

This means that, in order to determine the rate of speed at any instant of time, we substitute appropriately for t in this equation. At the end of 5 seconds, for example, the body is falling at the rate of 160 feet per second.

6.4 Exercises

1. If the total distance traveled is given by the equation $s = 16t^2$, determine the values for s from $t = 0$ to $t = 5$.
2. Average distance will be $16t^2/t$ or $16t$. Determine average values from $t = 0$ to $t = 5$.
3. We have seen that the rate of speed or marginal value is equal to $32t$. Determine marginal values from $t = 0$ to $t = 5$.
4. Plot total, average, and marginal curves on the same chart.
5. Draw a tangent to the total curve at $t = 3$. What is the slope of the tangent? How does this compare with the marginal value at $t = 3$?

6.5 The Techniques of Differentiation

It will be noted that when $s = 16t^2$, the derivative of this function is $ds/dt = 32t$. It will be found that in all cases the *coefficient* of the derivative function is equal to the coefficient of the original function multiplied by the exponent of the variable term in the original function; *i.e.*, $32 = 16 \times 2$; the *exponent* of the variable term in the derivative function is equal to the exponent of the original function minus one; *i.e.*, $2 - 1 = 1$.

In general, if

$$y = ax^n,$$

then

$$y' = na(x^{n-1}).$$

For example, if

$$y = 5x^4,$$
$$y' = 20x^3.$$

The following are the principal rules used in the differentiation of common functions:

Rule I. The derivative of the sum (or difference) of two functions is the sum (or difference) of the separate derivatives:

$$\frac{d}{dx}(u+v) = \frac{du}{dx} + \frac{dv}{dx},$$

and

$$\frac{d}{dx}(u-v) = \frac{du}{dx} - \frac{dv}{dx}.$$

Rule II. The derivative of the product of two functions is equal to the first function times the derivative of the second plus the second function times the derivative of the first:

$$\frac{d}{dx}(uv) = (u)\frac{dv}{dx} + (v)\frac{du}{dx}.$$

Rule III. The derivative of the quotient of two functions is equal to the denominator times the derivative of the numerator minus the numerator times the derivative of the denominator, all divided by the square of the denominator:

$$\frac{d}{dx}\left(\frac{u}{v}\right) = \frac{(v)\dfrac{du}{dx} - (u)\dfrac{dv}{dx}}{v^2}.$$

Rule IV. The derivative of a constant is equal to zero.

$$\frac{dK}{dx} = 0.$$

6.6 Exercises

Find the derivatives of the following functions:

1. $y=5x$.

2. $y=3/x$.

3. $y=x^3+7x^2-5x+10$.

4. $y=(1+x)^2$.

5. $y=1/(3x+2)$.

6. $y=mx+b$, where m and b are constants.

6.7 Bibliographical Note

The points made in this chapter are discussed in greater detail in any introductory calculus textbook. In addition the student may find the following helpful: R. G. D. Allen, *Mathematical Analysis for Economists*, pp. 134–148; W. L. Crum and Joseph A. Schumpeter, *Rudimentary Mathematics for Economists and Statisticians*, pp. 87–97; Irving Fisher, *A Brief Introduction to the Infinitesimal Calculus*, Introduction and Chapters 1–2; M. Richardson, *Fundamentals of Mathematics*, rev. ed., pp. 299–319; and Gerhard Tintner, *Mathematics and Statistics for Economists*, Chapter 10.

7

Use of Derivatives in Economic Analysis

7.1 Marginal as the Derivative of the Total

Among the most important relationships to the price theorist are the total, marginal, and average relationships. Marginal cost, for example, is the ratio of the change in total cost to a very small corresponding change in output. If we let t=total cost and x=output, this may be written:

$$MC = \frac{\Delta t}{\Delta x},$$

or, if we take the increment in output to be infinitesimally small,

$$MC = \frac{dt}{dx},$$

which is the derivative of total cost with respect to output. Or we may say that if

$$TC = f(x),$$

then

$$MC = f'(x).$$

Similarly, the average total cost is total cost divided by output. That is to say that if

$$TC = f(x),$$

then

$$ATC = \frac{f(x)}{x}.$$

Consider the total cost function:

$$TC = f(x) = ax^3 - bx^2 + cx + d.$$

The corresponding marginal cost function is

$$MC = f'(x) = 3ax^2 - 2bx + c,$$

and the corresponding average cost function is

$$ATC = \frac{f(x)}{x} = ax^2 - bx + c + \frac{d}{x}.$$

7.2 Exercises

1. Given $TC = x^3 - 4x^2 + 8x + 4$. Plot on the same chart from $x = 0$ to $x = 4$: (a) TC; (b) ATC; (c) MC.

2. Given $TR = 12x - 2x^2$. Plot on the same chart from $x = 0$ to $x = 4$: (a) TR; (b) AR; (c) MR.

3. Plot on the same chart ATC and MC from problem 1 above and AR and MR from problem 2 above.

4. Plot on the same chart TC from problem 1 above and TR from problem 2 above. Draw tangents to the TR and TC curves at $x = 2$.

5. A given average curve is a linear function passing through the points $P(0, 0)$ and $P'(8, 4)$. Plot this curve and the corresponding total and marginal curves on the same chart.

6. A given linear average curve passes through the point $P(6, 0)$ with a slope of 2. Draw the average, total, and marginal curves on the same chart.

7. Explain the material quoted below, which originally appeared in an article by J. R. Hicks in *Econometrica* in 1935 and was reprinted in *Readings in Price Theory*, edited by George J. Stigler and Kenneth E. Boulding, p. 363:

   "If the prices at which the monopolist hires his factors are fixed, his cost of production can be taken as a simple function of output. Let $F(x)$ be the total cost of producing an output of x.

   "If the monopolist's selling price is p, and $p = f(x)$ is the demand curve confronting him, his profit on selling output x will be

   $$xf(x) - F(x)$$

   which is maximized when

   $$xf'(x) + f(x) = F'(x).$$

"So much has been familiar since Cournot; the principal recent innovation has been to give the expression on the left of the last equation a name, *Marginal Revenue.* The equation can then be written

$$\text{Marginal Revenue} = \text{Marginal Cost}$$

which is certainly a convenient way of expressing the first condition of monopolistic equilibrium."

8. Explain the following analysis from Joan Robinson, *The Economics of Imperfect Competition*, p. 136:

"If A is the average cost, M the marginal cost, and O the output,

$$M = \frac{d(AO)}{dO}$$

$$= A+O\left(\frac{dA}{dO}\right)$$

$$M-A = O\left(\frac{dA}{dO}\right),$$

which is the increase in the cost of the old output, O, when output is increased by one unit."

7.3 Bibliographical Note

Discussions of marginal as the first derivative of the total will be found in R. G. D. Allen, *Mathematical Analysis for Economists*, pp. 152–157; W. L. Crum and Joseph A. Schumpeter, *Rudimentary Mathematics for Economists and Statisticians*, pp. 81–91; George J. Stigler, *The Theory of Price*, rev. ed., pp. 297–298; Gerhard Tintner, *Mathematics and Statistics for Economists*, pp. 88–92.

This is a good time for the student to read the path-breaking article by Jacob Viner, "Cost Curves and Supply Curves," *Zeitschrift für Nationalökonomie*, 1932, reprinted in Richard V. Clemence, ed., *Readings in Economic Analysis*, Vol. II, pp. 8–30, and George J. Stigler and Kenneth E. Boulding, ed., *Readings in Price Theory*, pp. 198–232.

8

Higher Derivatives

8.1 Higher Derivatives

It will be recalled that the first derivative by definition is the limit of $\Delta y/\Delta x$ as Δx approaches zero:

$$\frac{dy}{dx} = y' = f'(x) = \lim_{\Delta x \to 0} \frac{\Delta y}{\Delta x}.$$

The first derivative of y in terms of x will, in general, also be a function of x, and this function may be differentiated, giving the second derivative:

$$\frac{d^2y}{dx^2} = y'' = f''(x) = \lim_{\Delta x \to 0} \frac{\Delta y'}{\Delta x}.$$

The second derivative may in turn be differentiated, giving the third derivative:

$$\frac{d^3y}{dx^3} = y''' = f'''(x) = \lim_{\Delta x \to 0} \frac{\Delta y''}{\Delta x},$$

and so on.

If, for example,

$$y = f(x) = x^3 - 4x^2 + 8x + 4,$$

the first derivative is

$$y' = f'(x) = 3x^2 - 8x + 8,$$

the second derivative is

$$y'' = f''(x) = 6x - 8,$$

the third derivative is

$$y''' = f'''(x) = 6,$$

and the fourth derivative is

$$y'''' = f''''(x) = 0.$$

At $x=4$,

$$\begin{aligned} y &= f(4) = 36, \\ y' &= f'(4) = 24, \\ y'' &= f''(4) = 16, \\ y''' &= f'''(4) = 6, \\ y'''' &= f''''(4) = 0, \\ y^{v} &= f^{v}(4) = 0, \end{aligned}$$

and so on.

8.2 Exercises

1. Find the first three derivatives of the following functions:
   (a) $y=10-x^2$,
   (b) $y=3x^2-2x+3$,
   (c) $y=x^n$,
   (d) $y=ax^3+bx^2+cx+d$,
   (e) $p=100-5q$.
2. Find all of the derivatives of the function $y=f(x)=6x^6-5x^5+x^3+10$ that are not zero.

8.3 Interpretation of Higher Derivatives

As we have seen, the first derivative of a function $f(x)$ measures the rate of change of the function and the slope of the tangent to the curve $y=f(x)$ at the point in question. Similarly, the second derivative of a function $f(x)$ measures the rate of change of the first derivative of the function and the slope of the tangent to the curve $y'=f'(x)$ at the point in question.

If, for example, the total revenue function is assumed to be

$$T = y = f(x),$$

the marginal revenue function is the first derivative:

$$M = y' = f'(x).$$

If at $x=a$

$$y' = f'(a) > 0,$$

the total revenue function is increasing and the slope of the tangent to the total revenue curve at $x=a$ is positive. If at $x=b$

$$y' = f'(b) = 0,$$

the total revenue function is constant and the slope of the tangent to the total revenue curve at $x=b$ is zero. If at $x=c$

$$y' = f'(c) < 0,$$

the total revenue function is decreasing and the slope of the tangent to the total revenue curve at $x=c$ is negative.

Similarly, the second derivative measures the rate of change of the first derivative. If at $x=d$

$$y'' = f''(d) > 0,$$

marginal revenue is increasing at $x=d$. If at $x=e$

$$y'' = f''(e) = 0,$$

marginal revenue is constant at $x=e$. And if at $x=f$

$$y'' = f''(f) < 0,$$

marginal revenue is decreasing at $x=f$.

Consider the functions

$$\begin{aligned} y &= f(x) = \tfrac{5}{2}x^2 - \tfrac{1}{3}x^3 \\ y' &= f'(x) = 5x - x^2 \\ y'' &= f''(x) = 5 - 2x. \end{aligned}$$

These functions are plotted in Figure 8-1. It will be noted that from $x=0$ to $x=5$, the first derivative is greater than zero, and for these x-values $f(x)$ is increasing. When $f'(x)$ is zero, $f(x)$ is at a maximum. When $f'(x)$ is less than zero, $f(x)$ is decreasing.

From $x=0$ to $x=2\frac{1}{2}$, $f''(x)$ is greater than zero, and over this range $f'(x)$ is increasing. At $x=2\frac{1}{2}$, $f''(x)=0$ and $f'(x)$ is at a maximum. For values of x greater than $2\frac{1}{2}$, $f''(x)$ is less than zero and $f'(x)$ is decreasing.

From $x=0$ to $x=2\frac{1}{2}$, $f'(x)$ is positive and increasing, $f''(x)$ is positive, and $f(x)$ is increasing at an increasing rate. From $x=2\frac{1}{2}$ to $x=5$, $f'(x)$ is positive and decreasing, $f''(x)$ is negative, and $f(x)$ is increasing at a decreasing rate. We may generalize as follows:

If $f''(x)>0$, $f(x)$ is concave upward (convex from below).

If $f''(x)<0$, $f(x)$ is concave downward (convex from above).

At $f''(x)=0$, $f(x)$ is at a *point of inflection*, *i.e.*, where the curvature changes from convex to concave or from concave to convex.

8.4 Exercises

1. Given the total cost function:

$$T = y = f(x) = x^3 - 4x^2 + 8x + 4.$$

(a) What are the first four derivatives of this function?
(b) What is the marginal cost function?
(c) Plot $f(x)$, $f'(x)$, and $f''(x)$ on the same graph.
(d) What is the marginal cost at $x=3$?
(e) Draw a tangent to the total cost curve at $x=3$.
(f) What is the value of $f'(x)$ at $x=3$?
(g) What is the value of $f''(3)$?
(h) Draw a tangent to $f'(x)$ at $x=3$.
(i) What is the slope of the tangent to $f'(x)$ at $x=3$?
(j) State verbally the relation of $f'(x)$ to $f(x)$; of $f''(x)$ to $f'(x)$; of $f'''(x)$ to $f''(x)$.

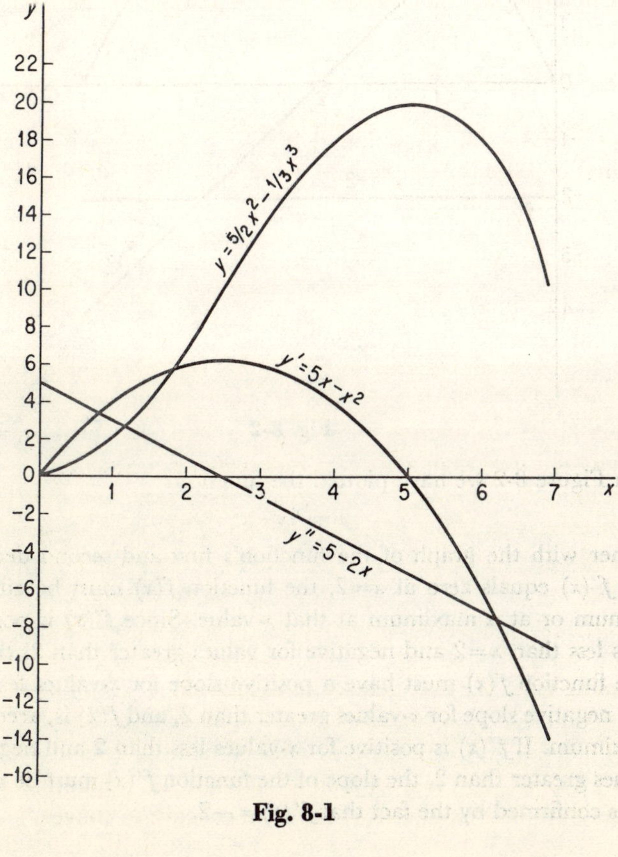

Fig. 8-1

8.5 Maxima and Minima

We have seen that when the function $f(x)$ is at a maximum (or minimum) value, *i.e.*, where the value of $f(x)$ is greater (or less) than all values in the immediate neighborhood of the point, the first derivative of the function is zero.

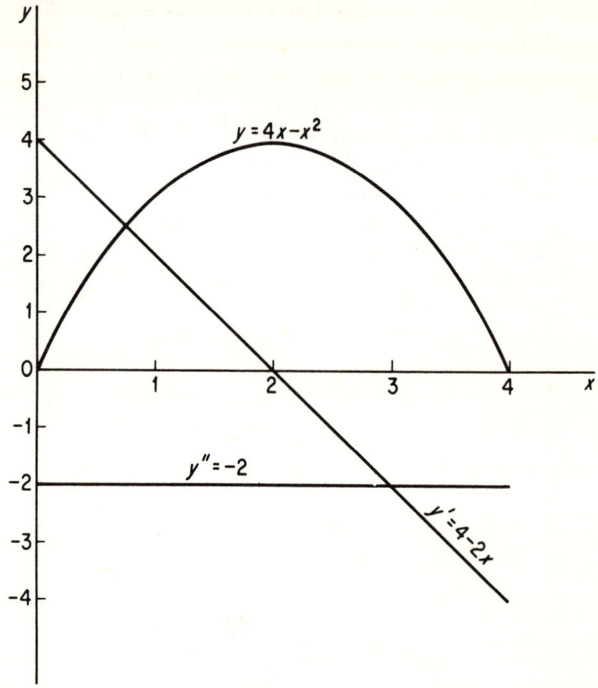

Fig. 8-2

In Figure 8-2 we have plotted the function

$$y = 4x - x^2$$

together with the graph of the function's first and second derivatives. Since $f'(x)$ equals zero at $x=2$, the function $f(x)$ must be either at a minimum or at a maximum at that x-value. Since $f'(x)$ is positive for values less than $x=2$ and negative for values greater than 2, the graph of the function $f(x)$ must have a positive slope for x-values less than 2 and a negative slope for x-values greater than 2, and $f(2)$ is, accordingly, a maximum. If $f'(x)$ is positive for x-values less than 2 and negative for x-values greater than 2, the slope of the function $f'(x)$ must be negative. This is confirmed by the fact that $f''(x) = -2$.

In Figure 8-3 we have plotted the function

$$y = x^2 - 4x + 6$$

together with the graph of the function's first and second derivatives. Since $f'(x)$ equals zero at $x=2$, the function $f(x)$ must be either at a maximum or at a minimum at that x-value. Since $f'(x)$ is negative for

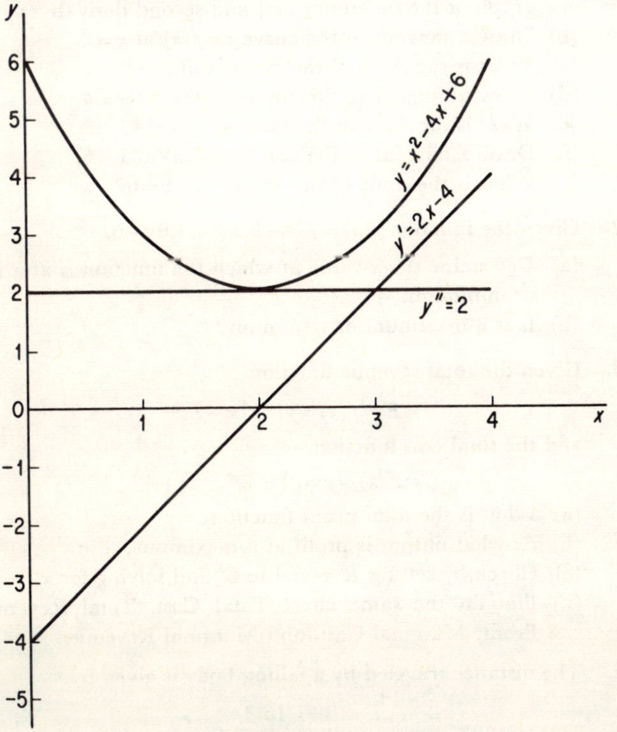

Fig. 8-3

values less than $x=2$ and positive for values greater than 2, the graph of the function $f(x)$ must have a negative slope for x-values less than 2 and a positive slope for x-values greater than 2, and $f(2)$ is, accordingly, a minimum. If $f'(x)$ is negative for x-values less than 2 and positive for x-values greater than 2, the slope of the function $f'(x)$ must be positive. This is confirmed by the fact that $f''(x) = 2$.

We may generalize as follows:

(1) All maximum and minimum values of the function $f(x)$ occur where $f'(x) = 0$.

(2) If $f'(a)=0$ and $f''(a)<0$, $f(a)$ is a maximum value of the function.

(3). If $f'(a)=0$ and $f''(a)>0$, $f(a)$ is a minimum value of the function.

8.6 Exercises

1. (a) Plot the function $f(x)=y=0.2x^2-1.6x+6$ together with the graph of the function's first and second derivatives.
(b) Draw a tangent to the curve $y=f(x)$ at $x=2$.
(c) What is the slope of the tangent at $x=2$?
(d) Draw a tangent to the curve $y=f(x)$ at $x=4$.
(e) What is the slope of the tangent at $x=4$?
(f) Draw a tangent to the curve $y=f(x)$ at $x=6$.
(g) What is the slope of the tangent at $x=6$?

2. Given the function $f(x)=y=-0.2x^2+1.6x+6$.

(a) Determine the x-value at which the function is at a maximum or minimum.
(b) Is it a maximum or minimum?

3. Given the total revenue function

$$R=f(x)=y=12x-2x^2$$

and the total cost function

$$C=F(x)=y=x^3-4x^2+8x+4.$$

(a) What is the total profit function?
(b) At what output is profit at a maximum?
(c) Check by setting R' equal to C' and solving for x.
(d) Plot on the same chart Total Cost, Total Revenue, Total Profit, Marginal Cost, and Marginal Revenue.

4. The distance traveled by a falling body is given by

$$s = 16t^2$$

where s is distance in feet and t is time in seconds.

(a) What are the first and second derivatives of this function?
(b) Express verbally in terms of speed and acceleration the meaning of the first and second derivatives of this function.

5. Determine the value (equal to, less than, or greater than zero) of $f'(a)$ and $f''(a)$ for each of the functions represented in Figure 8-4.

8.7 Bibliographical Note

A discussion of successive differentiation and maxima and minima will be found in any elementary calculus text. The following may also prove helpful: R. G. D. Allen, *Mathematical Analysis for Economists*, pp. 179–180; W. L. Crum and Joseph A. Schumpeter, *Rudimentary Mathematics for Economists*

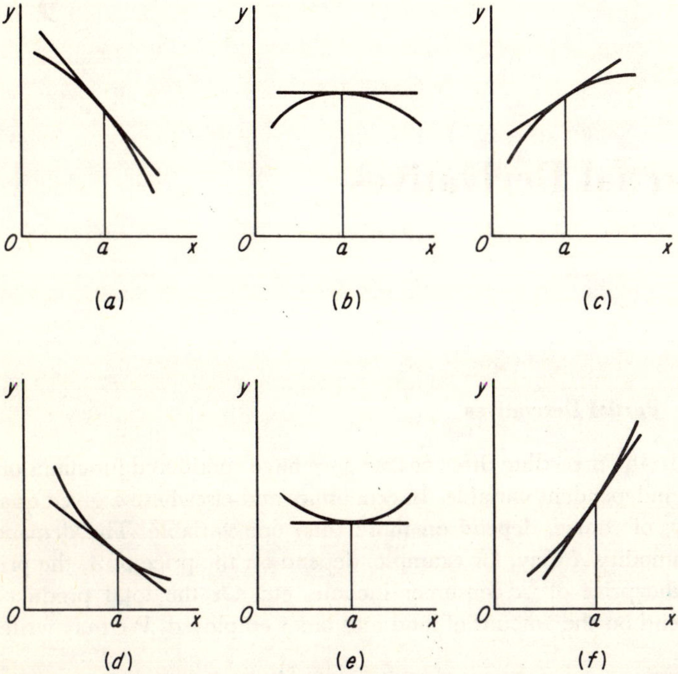

Fig. 8-4

and Statisticians, Chapter 5; Irving Fisher, *A Brief Introduction to the Infinitesimal Calculus*, Chapter 4; Gerhard Tintner, *Mathematics and Statistics for Economists*, Chapters 14 and 15.

9

Partial Derivatives

9.1 Partial Derivatives

In the preceding three chapters we have considered functions of only one independent variable. In economics and elsewhere a given quantity may, of course, depend on more than one variable. The demand for commodity A may, for example, depend on the price of A, the price of B, the price of C, consumer income, etc. Or the total product may depend on the amount of land and labor employed. We may write

$$z = f(x, y),$$

which means that for every permissible pair of values of the independent variables x and y there is at least one value of z that corresponds to this pair.

If we regard y as having some fixed value, $f(x, y)$ becomes a function of x alone, which is defined for all values of x in combination with the fixed value of y. Two functions can be obtained in this way: z as a function of x only (y fixed) and z as a function of y only (x fixed).

The derivatives of each of these functions can be determined, and they are called *partial derivatives*. One partial derivative is obtained when x is varied and y is held constant, and another partial derivative is obtained when y is varied and x is held constant. The partial derivative of z with respect to x may be indicated by any one of several symbols, the most common of which are $\partial z/\partial x$, $f(x, y)/\partial x$, $f_x(x, y)$, or z_x.

The partial derivative of z with respect to x is defined as

$$\frac{\partial z}{\partial x} = \frac{\partial f}{\partial x}(x, y) = \lim_{\Delta x \to 0} \frac{f(x+\Delta x, y) - f(x, y)}{\Delta x}.$$

Consider the function

$$z = 4x^2 - 2xy + 3xy^2 + y^3.$$

To find the partial derivative of z with respect to x, treat y as a constant and take the derivative with respect to x:

$$\frac{\partial z}{\partial x} = f_x(x, y) = 8x - 2y + 3y^2.$$

At the point $x=1, y=2$,

$$f_x(x, y) = 16.$$

Similarly, we can determine the partial derivative of z with respect to y, treating x as a constant:

$$\frac{\partial z}{\partial y} = f_y(x, y) = -2x + 6xy + 3y^2.$$

At the point $x=1, y=2$,

$$f_y(1, 2) = 22.$$

The idea of the partial derivative implies little that is new, and partial derivatives may be evaluated exactly as ordinary derivatives. The partial derivative of a function of a function, however, deserves special attention. If z is a single-valued function of u, and u in turn is a single-valued function of x and y, then z is a function of x and y with the partial derivatives

$$\frac{\partial z}{\partial x} = \frac{dz}{du} \cdot \frac{\partial u}{\partial x} \quad \text{and} \quad \frac{\partial z}{\partial y} = \frac{dz}{du} \cdot \frac{\partial u}{\partial y}.$$

In general, if $u = f(x, y)$ and $z = u^n$,

$$\frac{\partial z}{\partial x} = \frac{\partial u^n}{\partial x} = \frac{du^n}{du} \cdot \frac{\partial u}{\partial x} = nu^{n-1} \frac{\partial u}{\partial x}.$$

For example, if

$$u = 2x^2 + y^3$$

and

$$z = 4u^2,$$

then

$$\frac{\partial z}{\partial x} = 8u \cdot 4x = 32ux,$$

and

$$\frac{\partial z}{\partial y} = 8u \cdot 3y^2 = 24uy^2.$$

9.2 Exercises

1. Given the function $z=2x^2-3y+1$. Find both partial derivatives.
2. (a) Given the function $z=x^2+2xy-y^2$. Find both partial derivatives.

   (b) Evaluate them for $x=1$ and $y=2$.
3. Find both partial derivatives for the function $z=ax+by+c$.
4. Given the function $z=(x^2+y^2)^2$. Find both partial derivatives.
5. Given: $z=f(u)$ and $u=f(y/x)$. Find $\partial z/\partial x$.

9.3 Marginal Productivity

Suppose that two factors, land (A) and labor (B), are used in the production of the product (X). The production function is given by

$$x = f(a, b)$$

where x is the amount of the product, a is the amount of factor A employed, and b is the amount of factor B employed. If we assume that the amount of land (A) is fixed but that variable quantities of labor (B) may be employed, the marginal productivity of labor is given by

$$\frac{\partial x}{\partial b} = f_b(a, b).$$

Similarly, if the amount of labor is held constant but the amount of land is varied, the marginal productivity of land is given by

$$\frac{\partial x}{\partial a} = f_a(a, b).$$

If, for example, the production function is given by

$$x = 20a-10a^2+10ab,$$

the marginal productivity of A is

$$\frac{\partial x}{\partial a} = 20-20a+10b,$$

and the marginal productivity of B is

$$\frac{\partial x}{\partial b} = 10a.$$

At $a=2$ and $b=6$, the marginal productivity of A is 40, and the marginal productivity of B is 20.

9.4 Exercises

1. Suppose the production function is $x = a^2 + ab + b^3$.

(a) Determine the marginal productivity of A and B.

(b) Determine the marginal productivity of each factor when $a = 4$ and $b = 2$.

2. The production function of a commodity is $x = 20a + 40b + 8c - 2a^2 + 4b^2 + 6c^2 - abc^2$.

(a) Find the marginal productivities of A, B, and C.

(b) Determine these marginal productivities if $a = 1$, $b = 2$, and $c = 3$.

3. The production function for the U.S. during the period 1899–1922 was estimated by Paul Douglas to be $x = 1.01L^{0 \cdot 75}C^{0 \cdot 25}$, where x is total production, L is labor, and C is capital. Find the marginal productivities of L and C.

4. The production function for the U.S. from 1921–1941 was estimated by Gerhard Tintner to be $x = a^{0 \cdot 10}b^{0 \cdot 04}$, where x is total product, a is labor, and b is fixed capital. Find the marginal productivities of labor and capital.

9.5 Bibliographical Note

The technique for taking the partial derivative of a function is discussed in any calculus textbook, or the student may consult any of the following: R. G. D. Allen, *Mathematical Analysis for Economists*, pp. 296–314; D. W. Bushaw and R. W. Clower, *Introduction to Mathematical Economics*, pp. 230–232; W. L. Crum and Joseph A. Schumpeter, *Rudimentary Mathematics for Economists and Statisticians*, pp. 102–106; Irving Fisher, *A Brief Introduction to the Infinitesimal Calculus*, pp. 73–84; Gerhard Tintner, *Mathematics and Statistics for Economists*, pp. 145–151.

This is a good time for the student to review the general theory of production. Among the classics in this field are John M. Cassels, "On the Law of Variable Proportions," *Explorations in Economics*, reprinted in William Fellner and Bernard F. Haley, ed., *Readings in the Theory of Income Distribution*, pp. 103–118; Piero Sraffa, "The Laws of Returns under Competitive Conditions," *The Economic Journal*, 1926, reprinted in George J. Stigler and Kenneth E. Boulding, ed., *Readings in Price Theory*, pp. 180–198 and Richard V. Clemence, ed., *Readings in Economic Analysis*, Vol. II, pp. 54–69; Joan Robinson, *The Economics of Imperfect Competition*, Chapter 20; Frank H. Knight, *Risk, Uncertainty and Profit*, pp. 94–104. Textbook treatments of this topic will be found in M. M. Bober, *Intermediate Price and Income Theory*, Chapter 4; Richard H. Leftwich, *The Price System and Resource Allocation*, rev. ed., Chapter 7; George J. Stigler, *The Theory of Price*, rev. ed., pp. 111–121; Alfred W. Stonier and Douglas C. Hague, *A Textbook of Economic Theory*, pp. 72–82.

10

Homogeneous Functions

10.1 Homogeneity

If the variables x and y are increased or decreased in a fixed proportion, the corresponding increase or decrease in the function $z=f(x, y)$ may be either in equal, in greater, or in less proportion than the change in x and y. In the special case where $z=f(x, y)$ always increases in the same proportion as x and y, the function is said to be homogeneous of the first degree or to be linear and homogeneous. That is to say, if

$$f(kx, ky) = kf(x, y)$$

for any point (x, y) and for any value of k, the function $z=f(x, y)$ is a linear homogeneous function.

For example, if

$$z = 6x+4y,$$

then

$$\begin{aligned} kz &= 6kx+4ky \\ &= k(6x+4y). \end{aligned}$$

If the variables x and y are multiplied by k, z is multiplied by k, and the function $z=6x+4y$ is linear and homogeneous.

The linear homogeneous function is the simplest case of a wider class of homogeneous functions. If

$$f(kx, ky) = k^r f(x, y),$$

the function is homogeneous of the rth degree. In general, a function is said to be homogeneous of degree r if, when each of the independent variables is multiplied by a positive constant k, the new function is k^r times the original function.

Consider the function

$$z = f(x, y) = 6x^2+24xy+4y^2.$$

Then

$$\begin{aligned} f(kx, ky) &= 6k^2x^2+24kxky+4k^2y^2 \\ &= k^2(6x^2+24xy+4y^2). \end{aligned}$$

The value of r is 2, and the equation $z=6x^2+24xy+4y^2$ is homogeneous of degree 2.

Next consider the equation

$$\begin{aligned} z &= \frac{6x+4y}{5x+3y}. \\ kz &= \frac{6kx+4ky}{5kx+3ky} \\ &= \frac{k(6x+4y)}{k(5x+3y)} \\ &= \frac{6x+4y}{5x+3y}. \end{aligned}$$

In this case $r=0$; that is, $k^0=1$.

Particular examples of homogeneous functions follow:

(a) Homogeneous of zero degree:

$$z = \frac{a_1x+b_1y}{a_2x+b_2y}$$

(b) Homogeneous and linear:

(1) $z = ax+by$

(2) $z = \sqrt{ax^2+2hxy+by^2}$

(3) $z = ax^a y^{1-a}$

(4) $z = \dfrac{ax^2+2hxy+by^2}{cx+dy}$

(c) Homogeneous of the second degree:

$$z = ax^2+2hxy+by^2$$

(d) Homogeneous of degree $a+b$:

$$z = ax^a y^b$$

10.2 Exercises

1. Confirm each of the functions (a) to (d) in the preceding section for degree of homogeneity.

2. Show that the following are linear homogeneous functions:

   (a) $z = \sqrt[3]{x^2y}$;

   (b) $z = \sqrt{x^2+y^2}$;

   (c) $z = \dfrac{(x^3+y^3)}{(x^2+y^2)}$.

3. Determine the degree of homogeneity of the function
$$f(x, y) = x^3+6x^2y-3xy^2+y^3.$$

4. Determine the degree of homogeneity of the function
$$f(x, y, z) = 2x^2+4xy+z^2.$$
Check by multiplying the variables x, y, and z by 3.

5. What is the degree of homogeneity of the function
$$f(x, y, z, u) = 3x^2-2y^2-6z^2+4u^2?$$
Check by multiplying each of the variables by 2.

6. Suppose the price of commodity A is p_a and the price of B is p_b. The demand for A is $D_a = 50p_b/p_a$, and the demand for B is $D_b = 100p_a/p_b$.

   (a) Show that the two demand functions are homogeneous of zero degree.
   (b) Check by supposing that p_a and p_b are doubled.
   (c) Determine the demand for A and B if $p_a = 5$ and $p_b = 3$.
   (d) Show that the demand remains the same if the prices in (c) are tripled.
   (e) What conclusions can you draw concerning the demand for A and B if all prices change by the same per cent?

10.3 Euler's Theorem

If the function $z = f(x, y)$ is linear and homogeneous, then
$$f(kx, ky) = kf(x, y).$$
It can be shown that for such a function
$$z = x\cdot\frac{\partial z}{\partial x}+y\cdot\frac{\partial z}{\partial y}.$$
This identity is known as Euler's Theorem. The formal proof of this

proposition will not be given here, but its application can be illustrated.

Suppose that a production function is given by

$$z = 10x+5y.$$

This function is linear and homogeneous, and its partial derivatives are

$$\frac{\partial z}{\partial x} = 10$$

and

$$\frac{\partial z}{\partial y} = 5.$$

Then

$$\frac{x\partial z}{\partial x} + \frac{y\partial z}{\partial y} = 10x+5y = z.$$

If $\partial z/\partial x$ is the marginal product of factor X, $x(\partial z/\partial x)$ is the total payment to factor X, and if $\partial z/\partial y$ is the marginal product of factor Y, $y(\partial z/\partial y)$ is the total payment to factor Y, if every unit of each factor is paid its marginal product. The total product, accordingly, is allocated to the two factors of production on the basis of their marginal productivities.

Euler's Theorem may be stated more generally as follows: if $f(x, y)$ is a homogeneous function of the rth degree, then

$$xf_x(x,y)+yf_y(x,y) = rf(x,y).$$

Consider, for example, the second degree function

$$z = 5x^2+15xy+3y^2.$$

$$\begin{aligned} xf_x(x,y)+yf_y(x,y) &= x(10x+15y)+y(15x+6y) \\ &= 10x^2+15xy+15xy+6y^2 \\ &= 2(5x^2+15xy+3y^2) \\ &= 2z. \end{aligned}$$

10.4 Exercises

1. (a) Determine whether the function $z = 3x^4+6y^4$ is a homogeneous function.

(b) Apply Euler's Theorem.

2. Given the function $u = x+y+z$.

(a) Determine whether the function is homogeneous.

(b) Apply Euler's Theorem.

3. Given the production function $z=f(a, b)=3a^2+4ab+b^2$, where z is total product, a the amount of land employed, and b the amount of labor employed.

   (a) Determine whether the function is homogeneous.
   (b) Apply Euler's Theorem.
   (c) If $a=3$ and $b=5$, what is the total product?
   (d) If labor is paid on the basis of its marginal productivity, what is the wage rate?
   (e) What is the total amount paid to labor?
   (f) If land is the "residual claimant," what rent does land receive?

4. Determine whether the following function is homogeneous:

$$z = x^2+2xy+y^3.$$

10.5 Bibliographical Note

Discussions of homogeneous functions and Euler's Theorem will be found in R. G. D. Allen, *Mathematical Analysis for Economists*, pp. 315–322 and Gerhard Tintner, *Mathematics and Statistics for Economists*, Chapter 17. The economic implications of these concepts are discussed in M. M. Bober, *Intermediate Price and Income Theory*, pp. 377–390; Joan Robinson, *Collected Economic Papers*, pp. 1–19; and Alfred W. Stonier and Douglas C. Hague, *A Textbook of Economic Theory*, pp. 328–342. An interesting effort to explain a linear and homogeneous production function in strictly nonmathematical terms will be found in G. L. S. Shackle, *Economics for Pleasure*, Chapter 20.

11

Elasticity of Demand and Supply

11.1 The Concept of Elasticity

We know that the demand function can be written $y=f(x)$, where y=price and x=quantity demanded. We also know that in general the functional relationship between price and quantity demanded is an inverse one; that is, the lower the price, the greater the quantity demanded. But a given change in price may cause the quantity of one commodity demanded to change a great deal, while the same price change in another commodity may have but a slight effect on the quantity demanded. As a device for measuring the responsiveness of demand to changes in price, the concept of demand elasticity was invented. This tool may also be applied to supply, cost, and a great many other concepts used by the economist. We will have more to say about elasticity as it applies to these ideas in subsequent chapters.

The elasticity of the function $y=f(x)$ with respect to y may be represented by the symbol Ex/Ey, and it is defined as the limit of the ratio of the relative increment in x to the relative increment in y as y approaches zero:

$$\frac{Ex}{Ey} = \lim_{\Delta y \to 0} \frac{(\Delta x/x)}{(\Delta y/y)} = \frac{dx/x}{dy/y} = \left(\frac{dx}{x}\right)\cdot\left(\frac{y}{dy}\right) = \left(\frac{y}{x}\right)\cdot\left(\frac{dx}{dy}\right) = \frac{y}{x}\left(\frac{1}{dy/dx}\right).$$

Consider the demand function $y=25-5x$.

$$\frac{Ex}{Ey} = \frac{y}{x}\left(\frac{1}{dy/dx}\right) = \left(\frac{25-5x}{x}\right)\frac{1}{-5} = \frac{25-5x}{-5x} = \frac{5-x}{-x}.$$

At $x=2\frac{1}{2}$, the elasticity of demand with respect to price is

$$\frac{5-2\frac{1}{2}}{-2\frac{1}{2}} = -1.$$

To determine the elasticity of demand at any given price, substitute for y in the equation of the curve; solve for x; and then substitute for x in the formula for elasticity. For example, to determine the elasticity of demand at $y=12\frac{1}{2}$ in the demand function $y=25-5x$:

$$12\tfrac{1}{2} = 25-5x$$
$$x = 2\tfrac{1}{2}$$
$$\frac{Ex}{Ey} = \frac{5-2\frac{1}{2}}{-2\frac{1}{2}} = -1.$$

11.2 Exercises

1. Find the elasticity of demand with respect to price at $D=2$ when $p=10-3D$.

2. $x+y-10=0$ is a demand function where x is quantity demanded and y is price. Find the elasticity of demand with respect to price at $y=6$.

3. The total revenue function is $TR=x^2+4xy-100x$. What is the elasticity of demand at $x=12$?

4. Prove that when the demand function is given in the general form $Ax+By-C=0$, the elasticity of demand with respect to price is $1-(C/Ax)$, where $C/A=a=$ the x-intercept of the demand curve; *i.e.*, $E_d=1-a/x$.

5. What is the elasticity of demand when $y=a$, if a is a constant?

6. What is the elasticity of demand when $xy=K$, if K is a constant?

7. What is the elasticity of supply when $p=5S$?

8. (a) What is the elasticity of supply when $y=2x^3$?
   (b) Prove that if $y=ax^n$, $E_s=1/n$.

9. What is the elasticity of supply when $x=5y$?

10. (a) What is the elasticity of supply at $x=2$ if $x-y+10=0$?
    (b) What is the elasticity of supply at $x=2{,}000$?

11. Prove that when the supply function is given in the general form $Ax-By+C=0$, the elasticity of supply with respect to price is $(C/Ax)+1$; *i.e.*, $E_s=1-a/x$ since $a=-C/A$.

12. Show that for any linear supply curve:

(a) If $a=0$, $e=1$, where a is the x-intercept of the curve and e is elasticity of supply;

(b) If $a>0$, $e<1$;

(c) If $a<0$, $e>1$.

11.3 Partial Elasticities of Demand

If the demand for commodity A depends on the price of A and the price of B, we may write $D_a=f(p_a, p_b)$. The partial elasticity of demand for A with respect to the price of A is defined as the limit of the ratio of the relative change in demand $(\Delta D_a/D_a)$ to the relative change in price $(\Delta p_a/p_a)$ as $\Delta p_a \to 0_x$ and p_b remains fixed:

$$\frac{ED_a}{Ep_a} = \lim_{\Delta p_a \to 0} \frac{\Delta D_a/D_a}{\Delta p_a/p_a} = \left(\frac{p_a}{D_a}\right)\cdot\left(\frac{\partial D_a}{\partial p_a}\right).$$

Similarly,

$$\frac{ED_a}{Ep_b} = \left(\frac{p_b}{D_a}\right)\cdot\left(\frac{\partial D_a}{\partial p_b}\right).$$

The partial elasticity of the demand for A with respect to the price of A is approximately the per cent increase or decrease in the amount of A demanded if the price of A increases by 1 per cent while the price of B is constant. The partial elasticity of the demand for A with respect to the price of B is approximately the per cent increase or decrease in the amount of A demanded if the price of B increases by 1 per cent while the price of A is constant. This is also known commonly as the cross elasticity of demand for A with respect to the price of B.

Suppose that the demand function for commodity A is

$$D_a = 100-10p_a+5p_b.$$

The partial elasticity of demand for A with respect to the price of A is

$$\frac{ED_a}{Ep_a} = \frac{-10p_a}{100-10p_a+5p_b}$$

and

$$\frac{ED_a}{Ep_b} = \frac{5p_b}{100-10p_a+5p_b}.$$

At $p_a=10$ and $p_b=5$,

$$\frac{ED_a}{Ep_a} = \frac{-100}{25} = -4$$

and

$$\frac{ED_a}{Ep_b} = \frac{25}{25} = 1.$$

11.4 Exercises

1. The demand for commodity A is $D_a = 60 - 5p_a + 10p_b$.
   (a) Find the partial elasticities ED_a/Ep_a and ED_a/Ep_b.
   (b) Evaluate the partial elasticities of demand for $p_a = 10$ and $p_b = 5$.

2. If $z = x^a y^b$ where a and b are constants, determine the partial elasticities Ez/Ex and Ez/Ey.

3. Henry Schultz found $D_a = 63.3 - 1.9p_a + 0.2p_b + 0.5p_c$, where a is beef, b is pork, and c is mutton. Find the partial elasticities of the demand for beef with respect to the price of beef, the price of pork, and the price of mutton.

4. Let A be margarine and B be butter. W. R. Pabst has estimated the demand for margarine in the U.S., 1904–1933, as

$$D_a = p_a^{-1.32} p_b^{0.4}.$$

   Find the partial elasticities of demand.

5. Let x and p represent the demand and price for margarine and y and q the demand and price for butter. Suppose that

$$x = p^{-1.3} q^{0.4} \quad \text{and} \quad y = p^{0.3} q^{-0.1}.$$

   Find the four partial elasticities.

6. The partial elasticity of demand turns out to be sometimes positive and sometimes negative. What is the significance of the algebraic sign of the partial derivative of the demand for A with respect to the price of B?

11.5 Bibliographical Note

The student would do well at this point to read standard treatments of the concept of elasticity of demand. Two classics in this area are Alfred Marshall, *Principles of Economics*, 8th ed., Book III, especially Chapter 4; and Henry Schultz, *The Theory and Measurement of Demand*, Chapters 2–6. Textbook treatments that may prove helpful include R. G. D. Allen, *Mathematical Analysis for Economists*, pp. 251–257; Clark Lee Allen, James M. Buchanan, and Marshall R. Colberg, *Prices, Income, and Public Policy*, 2nd ed., pp. 20–26; Joe S. Bain, *Price Theory*, pp. 40–56; Paul H. Daus and William M. Whyburn, *Introduction to Mathematical Analysis*, pp. 140–141; Ralph K. Davidson, Vernon L. Smith, and Jay W. Wiley, *Economics: an Analytical Approach*, Chapter 2; George J. Stigler, *The Theory of Price*, rev. ed., Chapter 4; Alfred W. Stonier and Douglas C. Hague, *A Textbook of Economic Theory*, pp. 9–24; Sidney Weintraub, *Price Theory*, pp. 31–43; Gerhard Tintner, *Mathematics and Statistics for Economists*, pp. 118–120 and 151–152.

12

Elasticity of Demand and Marginal Revenue

12.1 Relationship

There is a precise relationship between the elasticity of demand and marginal revenue which serves the economist as a useful tool. Let p=price, x=quantity demanded, and e=elasticity of demand with respect to price. The demand curve is $p=f(x)$, and total revenue is $R=p\cdot x$. It follows from the rule for the derivative of a product that marginal revenue is

$$R' = \left(\frac{dp}{dx}\right) x+p.$$

The elasticity of demand with respect to price is

$$e = \frac{dx/x}{dp/p} = \frac{dx}{x}\cdot\frac{p}{dp} = \frac{p}{x}\cdot\frac{dx}{dp} = \frac{p}{x}\left(\frac{1}{dp/dx}\right).$$

From the formula for elasticity it follows that

$$e\left(\frac{dp}{dx}\right) = \frac{p}{x}$$

$$\frac{dp}{dx} = \frac{p}{ex}.$$

By substituting this in the formula for marginal revenue we get

$$R' = \left(\frac{p}{ex}\right) x+p = p\left(\frac{1}{e}+1\right).$$

This is to say that marginal revenue is equal to price times 1 plus the reciprocal of elasticity.

To check the derivation of marginal revenue in terms of elasticity in this way, consider for example the demand function $p=25-5x$. Then

$$R = p\cdot x = 25x-5x^2,$$
$$R' = f'(x) = 25-10x,$$

and

$$e = \left(\frac{25-5x}{x}\right)\frac{1}{-5} = \frac{5-x}{-x}.$$

Substituting this value for elasticity in the formula for marginal revenue, we get

$$R' = p\left(\frac{1}{e}+1\right) = p\left(\frac{-x}{5-x}+1\right).$$

Substituting for p:

$$R' = (25-5x)\cdot\left(\frac{-x}{5-x}+1\right) = \frac{-25x+5x^2}{5-x}+25-5x$$
$$= -5x+25-5x = 25-10x.$$

This is the same value for marginal revenue that we got directly by differentiating the total revenue function.

12.2 Exercises

1. State the formula $R'=p(1+1/e)$:
   (a) Explicitly in terms of e;
   (b) Explicitly in terms of p.

2. Given the demand curve $p=1{,}000-2D-3D^2$. Find:
   (a) R;
   (b) e;
   (c) R'.

3. Given the demand curve $y=(1-x)\ (2-3x)$. Find:
   (a) R;
   (b) e;
   (c) R'.

4. What is R when $e=-1$ for all values of x?

5. What is R' when $e=$infinity for all values of x?

6. What is e when $R'=0$?

7. What is e when $p=R'$?

8. Given the total revenue function $xy=K$, where K is a constant. Find:

(a) e;
(b) R'.

9. If price is \$2 and marginal revenue is \$1.40, what is the elasticity of demand?

10. If elasticity of demand is -0.9 and price is \$4, what is the marginal revenue?

11. If the ratio of price to marginal revenue is 9:4, what is the elasticity of demand?

12. Assume a firm that is trying to maximize profits uses a 100 per cent markup over marginal cost to determine its selling price. What implicit assumption does the management of the firm make with reference to elasticity of demand?

12.3 Bibliographical Note

Brief discussions of the relationship between elasticity of demand and marginal revenue will be found in the following: R. G. D. Allen, *Mathematical Analysis for Economists*, pp. 257–260; Ralph K. Davidson, Vernon L. Smith, and Jay W. Wiley, *Economics: an Analytical Approach*, pp. 21–22; Joseph P. McKenna, *Intermediate Economic Theory*, pp. 22–26; Alfred W. Stonier and Douglas C. Hague, *A Textbook of Economic Theory*, pp. 90–99; Gerhard Tintner, *Mathematics and Statistics for Economists*, pp. 121–122; Sidney Weintraub, *Price Theory*, pp. 24–44.

13

Total as the Integral of the Marginal

13.1 The Process of Integration

We have seen that if we have a total function given, we can derive the corresponding marginal function by taking the first derivative of the total function. But we may upon occasion have the marginal function given and need to derive from it the corresponding total function. The mathematical process by which this is done is called integration, and this comprises the second half of the calculus. Integration is the inverse process of differentiation in much the same way that division is the inverse process of multiplication. In differential calculus we know, for example, that if

$$y = 3x^2 + 4x,$$

then

$$\frac{dy}{dx} = 6x + 4$$

and the differential of y is

$$dy = (6x+4)\,dx.$$

In integral calculus we start with the function $dy = (6x+4)\,dx$ and work back to the function $y = 3x^2 + 4x$. This is written:

$$\int (6x+4)\,dx = 3x^2 + 4x.$$

But since $(d/dx)\ (3x^2+4x+C)$ also equals $6x+4$, where C is any constant, we must add an arbitrary constant to the integral:

$$\int (6x+4)\ dx = 3x^2+4x+C.$$

The principal general rules for integration follow:

$$\int dx = x+C.$$

$$\int ax^n\ dx = \frac{ax^{n+1}}{n+1}+C.$$

13.2 Exercises

Evaluate the following integrals, checking the answer in each case by differentiation:

1. $\int x^3\ dx.$ **2.** $\int (x-3)\ dx.$ **3.** $\int (1-2x)\ dx.$

4. $\int (t^2-3t)\ dt.$ **5.** $\int \frac{dx}{x^2}.$ **6.** $\int (1-y)^2\ dy.$

13.3 The Integral as the Area under a Curve

One of the most important uses of the integral is to determine the area under a nonlinear curve. In Figure 13-1 suppose that the curve AB represents the function $y=f(x)$. Then assume an increment of $x=\Delta x=SV=MN$, and consider the resulting increment, not of y, but of the area $OASM$ or z. This increment Δz is the small area $MSRN$, which is the sum of the rectangle $MSVN$ and the little triangular area SRV. The area of the rectangle is the product of the base, Δx, and its altitude, $f(x)$. It follows that

$$\Delta z = f(x)\Delta x+SRV.$$

The smaller we take Δx, the smaller the little triangle becomes relative to the rectangle, and as Δx approaches zero, the triangle approaches zero. We may write

$$\Delta z = f(x)\Delta x+SRV,$$

$$\frac{dz}{dx} = f(x),$$

$$dz = f(x)\ dx,$$

$$z = \int f(x)\ dx = \int ydx.$$

13.4 Example

Suppose that the equation of a curve is $y=2x^2+4$, and we wish to find the area, z, under the curve in terms of x from $x=0$. We know that

$$dz = (2x^2+4)\,dx$$

$$z = \int (2x^2+4)\,dx$$

$$z = \frac{2x^3}{3}+4x+C.$$

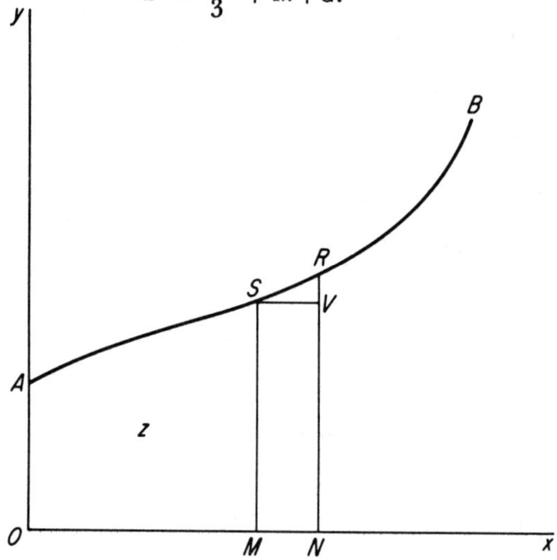

Fig. 13-1

In this problem we can also determine the value of C. We are measuring the area z from the y-axis. As x approaches zero, z also approaches zero. If we substitute zero for x and z in the equation

$$z = \frac{2x^3}{3}+4x+C,$$

we get

$$0 = 0+0+C,$$

so that

$$C = 0.$$

Therefore,

$$z = \frac{2x^3}{3}+4x.$$

At, for example,

$$x = 3,$$
$$z = 18+12 = 30.$$

13.5 Exercises

1. Plot the function $y=2x+4$. Determine the area z under the curve from $x=0$ to $x=4$ by simple arithmetic. Confirm this result by evaluating $z=\int (2x+4)\,dx$.

2. Suppose it is known that marginal cost $= 3x^2-8x+10$ and fixed cost $= 4$. Find
   (a) Total cost.
   (b) Total variable cost.
   (c) Average total cost.
   (d) Average variable cost.
   (e) Average fixed cost.

3. Marginal revenue equals $12-4x$.
   (a) Find average revenue.
   (b) At what quantity is elasticity of demand equal to -1?

4. Marginal utility $= 10x-6x^2$. Find total utility at $x = 2$.

5. Is it accurate to say that the area under a marginal revenue curve equals total revenue, and the area under a marginal utility curve equals total utility, but the area under a marginal cost curve does not equal total cost? Explain.

13.6 Bibliographical Note

The process of integration is, of course, described in any elementary calculus textbook. In addition, the following may prove helpful: R. G. D. Allen, *Mathematical Analysis for Economists*, Chapter 15; Paul H. Daus and William M. Whyburn, *Introduction to Mathematical Analysis*, Chapter 4; Irving Fisher, *A Brief Introduction to the Infinitesimal Calculus*, Chapter 6; M. Richardson, *Fundamentals of Mathematics*, rev. ed., pp. 328–337; Gerhard Tintner, *Mathematics and Statistics for Economists*, Chapter 19.

13.7 Review Problem

Before beginning a consideration of the geometry of price theory, which will commence with the next chapter, the following review problem may prove helpful:

Given:

$$\text{marginal cost} = MC = 3x^2-8x+8,$$
$$\text{fixed cost} = FC \doteq 4,$$
$$\text{marginal revenue} = MR = 12-4x.$$

Find:

1. TC;
2. ATC;
3. VC;
4. AVC;
5. AFC;
6. TR;
7. AR;
8. profit;
9. output where AVC is at a minimum;
10. output where TR is at a maximum;
11. output where MC is at a minimum;
12. output where TC is at a point of inflection;
13. most profitable output;
14. TC at most profitable output;
15. VC at most profitable output;
16. MC at most profitable output;
17. AR at most profitable output;
18. MR at most profitable output;
19. ATC at most profitable output;
20. AVC at most profitable output;
21. TR at most profitable output;
22. total profit at most profitable output;
23. elasticity of AR at most profitable output;
24. elasticity of MC at most profitable output.

Now suppose that the cost and revenue functions are given in the following form:

$$MC = ax^2 - bx + c,$$
$$FC = d,$$
$$MR = A - Bx.$$

Answer the first thirteen questions above.

C

Geometry

14

Geometry of Average-Marginal Relationships

14.1 Derivation of Marginal from Average

With this chapter we begin the analysis of price theory with the assistance of the tools of plane geometry. The solution of problems geometrically is both easier and more difficult than what we have been doing in previous chapters. It is easier in that those with limited mathematical backgrounds can follow the analysis without too great difficulty, and for this reason geometric charts are usually included in textbooks in elementary economics where no background in college mathematics is assumed. But the student will discover that in working out solutions to economic problems by geometrical methods he encounters difficulties that would disappear if the more powerful tools of calculus were employed. To arrive at a geometric solution often requires ingenuity of a high order, and frequently the process of trial and error appears to be the only way. The student will likely feel that the arrangement of topics in this book is essentially in order of increasing difficulty.

One should be careful not to underestimate the geometric technique in economic analysis. For those with minimal backgrounds in mathematics, the geometric method is the only mathematical technique available to them, and most economists spend a considerable fraction of their time trying to explain economic matters to those with limited mathematical backgrounds. But even for those whose mathematical training has been more comprehensive, the ability to present ideas in a two-dimensional figure often provides insights that would otherwise be lacking. And geometry will be more meaningful to those who have some knowledge of the calculus.

Among the most important relationships in price theory are the

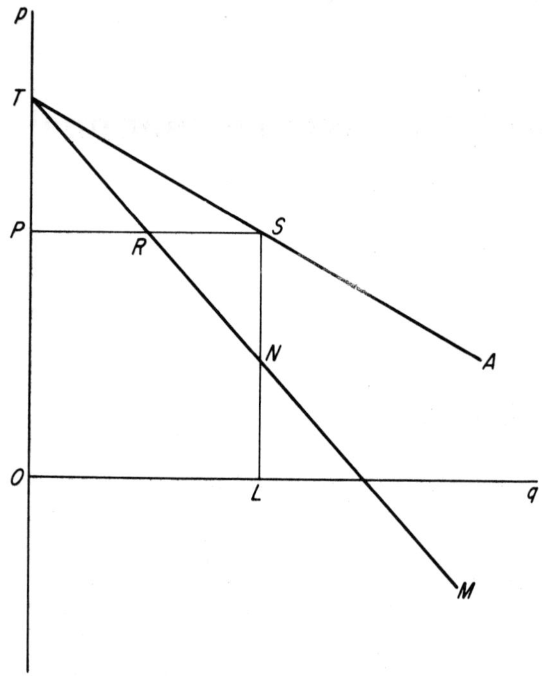

Fig. 14-1

average-marginal relationships. In Figure 14-1, A is an average curve and M is the corresponding marginal curve. Then

$$\text{Total} = OP \times OL = OPSL$$

and

$$\text{Total} = \text{area under marginal curve} = OTNL.$$

Therefore,

$$OPSL = OTNL,$$
$$OPRNL + RSN = OPRNL + PTR,$$

and

$$RSN = PTR \text{ in area.}$$

But

$$\angle TPR = \angle RSN$$

and

$$\angle PRT = \angle SRN;$$

therefore,

$$\angle PTR = \angle RNS,$$
$$\triangle RSN \cong \triangle PTR,$$

and

$$PR = RS.$$

We may generalize: when the average curve is linear, the corresponding marginal curve will have the same y-intercept as the average curve and will bisect any horizontal line drawn from the y-axis to the average curve. In Figure 14-1, if average $= OP = LS$, marginal $= LN$.

14.2 Exercises

1. Draw a positively inclined linear average curve and prove that the corresponding marginal curve has the same intercept on the y-axis as the average curve and bisects any horizontal line drawn from the y-axis to the average curve.

2. If the distance from the origin to the x-intercept of a linear demand curve is OM, what is the distance from the origin to the x-intercept of the corresponding marginal revenue curve?

3. Draw a nonlinear demand curve and determine the marginal revenue at three x-values. Sketch in the corresponding marginal curve.

4. What is the marginal revenue curve corresponding to a horizontal demand curve?

5. What is the marginal revenue curve corresponding to a vertical demand curve?

6. Draw the marginal revenue curve corresponding to the kinked demand curve in Figure 14-2.

7. In Figure 14-3 there are four demand curves: AB, CD, ALD, and CLB.
   (a) What is the marginal revenue curve corresponding to each demand curve?
   (b) If the demand curve is ALD and the marginal cost curve is MC, what will be the most profitable output?

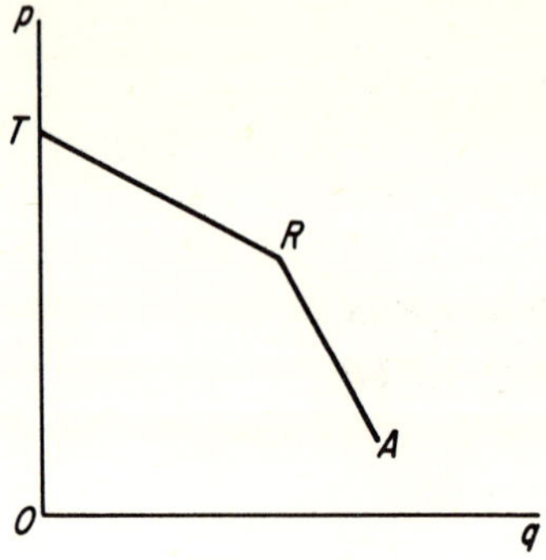

Fig. 14-2

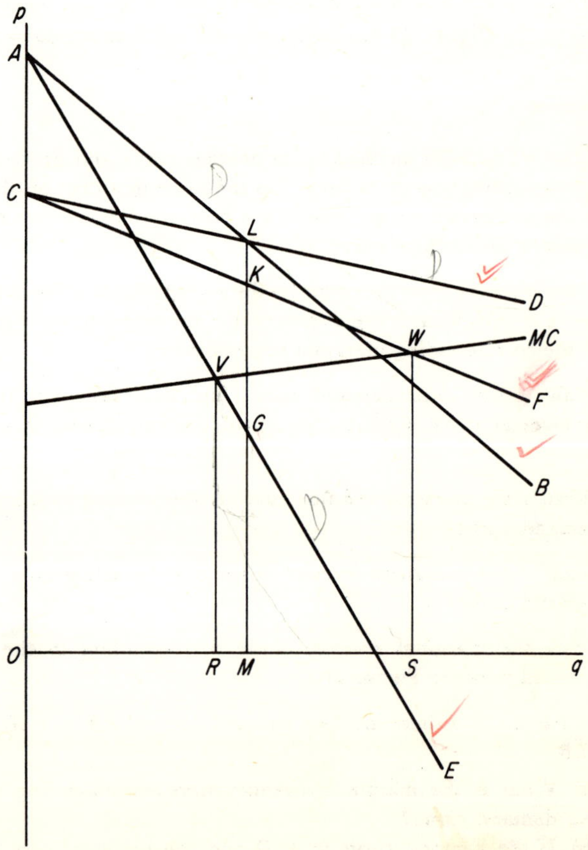

Fig. 14-3

8. Show that if two linear average curves intersect, the point of intersection of the two corresponding marginal curves will have the same y-value as the point of intersection of the average curves.

9. Draw a U-shaped average curve and sketch in as accurately as possible the corresponding marginal curve.

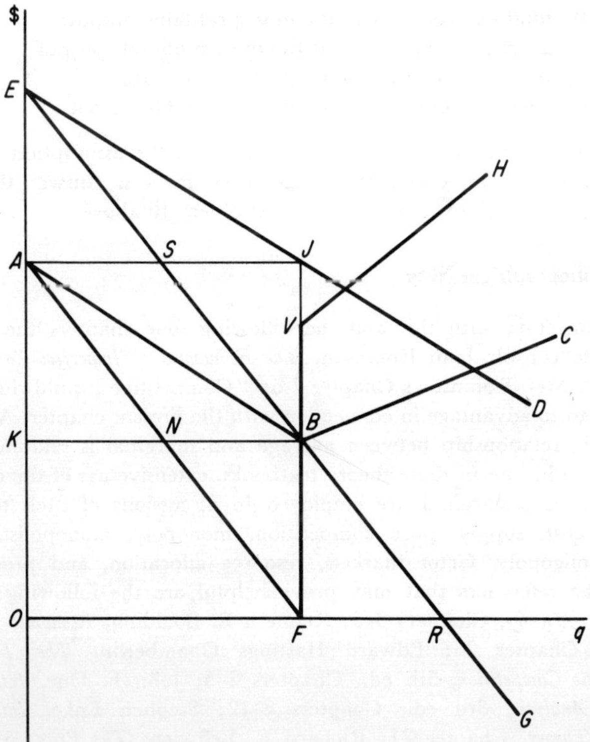

Fig. 14-4

10. In Figure 14-4 costs and revenues are represented by straight lines. *ABC* is the average total cost curve, and *ED* is the demand curve.

Determine the following:

(a) the marginal cost curve;
(b) the marginal revenue curve;
(c) the most profitable output;
(d) the price at which the most profitable output can be sold;

(e) the elasticity of demand at the most profitable output;
(f) marginal revenue at the most profitable output;
(g) output at which total revenue is at a maximum;
(h) total revenue at the most profitable output;
(i) total cost at the most profitable output;
(j) total profit at the most profitable output;
(k) average total cost at the most profitable output;
(l) total variable cost at the most profitable output;
(m) average variable cost at the most profitable output;
(n) total fixed cost at the most profitable output;
(o) average fixed cost at the most profitable output.

11. Draw a figure similar to Figure 14-4 on the assumption that the firm sells in a perfectly competitive market. Answer the same questions about this figure as in Problem 10 above.

14.3 Bibliographical Note

In connection with this and the following four chapters the student is urged to consult Joan Robinson, *The Economics of Imperfect Competition*, Chapter 2. Mrs. Robinson's Chapter 7 on "Competitive Equilibrium" can also be read to advantage in connection with the present chapter. Although the specific relationship between average and marginal is seldom spelled out in detail in intermediate theory textbooks, extensive use of the concepts of average and marginal are employed in discussions of such topics as demand, cost, supply, pure competition, monopoly, monopolistic competition, oligopoly, factor markets, resource allocation, and production. Among the references that may prove helpful are the following: Joe S. Bain, *Price Theory*, Chapters 3–6; Kenneth E. Boulding, *Economic Analysis*, 3rd ed., Chapter 25; Edward Hastings Chamberlin, *The Theory of Monopolistic Competition*, 5th ed., Chapters 2–5; John F. Due, *Intermediate Economic Analysis*, 3rd ed., Chapters 8–12; Stephen Enke, *Intermediate Economic Theory*, Chapter 21; Richard F. Leftwich, *The Price System and Resource Allocation*, rev. ed., Chapters 9–14; Abba P. Lerner, *The Economics of Control*, Chapter 7; Joseph P. McKenna, *Intermediate Economic Theory*, Chapters 2, 5, 9–12; George J. Stigler, *The Theory of Price*, rev. ed., Chapters 10–12; Sidney Weintraub, *Price Theory*, Chapters 2, 5, 6.

15

Total-Average and Total-Marginal Relationships

15.1 Derivation of Average from Total

In Figure 15-1, *TP* is total product. If *OM* units of input are employed, the total product will be *MV*. Average product is total product

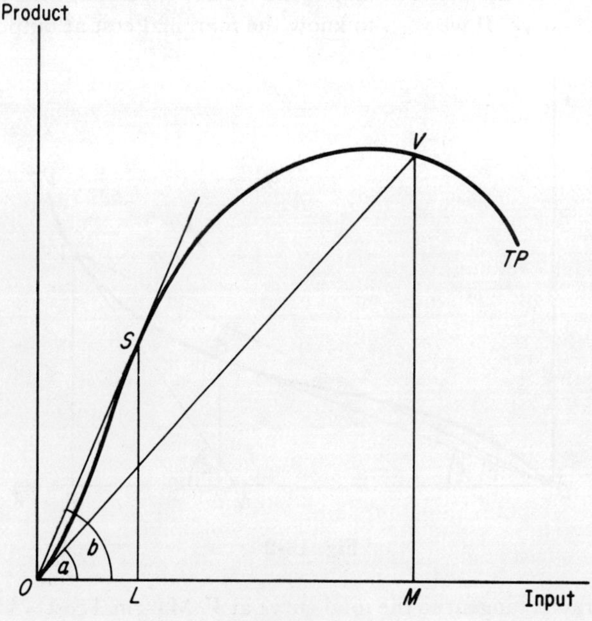

Fig. 15-1

divided by input, *i.e.*, $MV \div OM$. But from the point of view of angle a, $MV \div OM$ is the opposite side of the triangle OMV divided by the adjacent side, which by definition is the tangent of angle a and is written "tan a." We may generalize by saying that to determine average product for any input, OM, draw a vertical line from M to the total curve at V; draw a line from V to the origin. The tangent of the angle formed by OV and the x-axis is the average product.

Since the larger the angle, the greater the tangent of the angle, average product will be greatest at input OL. The line OS is tangent to the curve at S; if the angle b were any larger, the line OS would miss the curve entirely. We may generalize by saying that the input where average product is greatest may be determined by drawing a line through the origin and tangent to the total product curve; the point of tangency determines the input where average product is greatest. In Figure 15-1, average product is greatest at input OL and is equal to tan b.

15.2 Derivation of Marginal from Total

In Figure 15-2, VC is the total variable cost curve. We have seen that marginal is the rate of change of the total or the slope of the tangent to the total curve. If we wish to know the marginal cost at output OM,

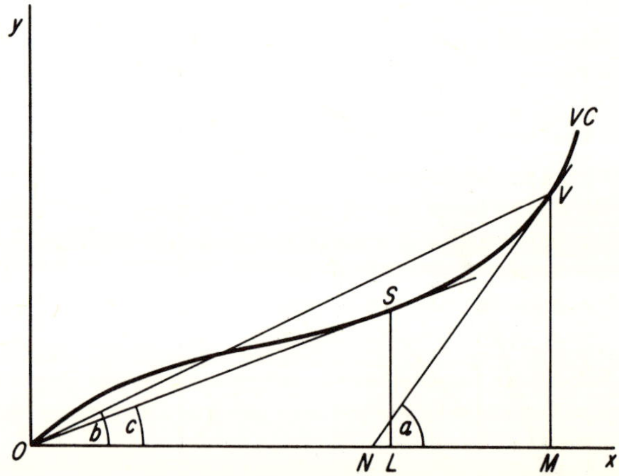

Fig. 15-2

we may draw a tangent to the total curve at V. Marginal cost $= MV/NM = \tan a$. Average variable cost at this output is $MV/OM = \tan b$.

If we draw a tangent to the total curve so that the tangent passes through the origin (*OS* in the figure), we have located the output where average variable cost is at a minimum; tan *c* represents the lowest possible average variable cost. But since *OS* is tangent to the total curve, tan *c* also measures marginal cost at output *OL*. We may generalize by saying that when average variable cost is at a minimum, marginal cost equals average variable cost.

15.3 Exercises

1. Demonstrate that the greater the angle is, the greater the tangent of the angle.
2. Draw a bell-shaped total product curve and sketch in the corresponding average product and marginal product curves. Demonstrate each of the following propositions:
   (a) When average is increasing, marginal is greater than average.
   (b) When average is decreasing, marginal is less than average.
   (c) When average is at a maximum (or minimum), marginal equals average.
   (d) When total is increasing, marginal is positive.
   (e) When total is decreasing, marginal is negative.
   (f) When total is at a maximum (or minimum), marginal is zero.
   (g) When marginal is at a maximum (or minimum), total is at a point of inflection.
3. Assume that a firm sells in a perfectly competitive market. Prepare a two-part chart.
   (a) In the upper half draw a total revenue and total cost curve. Indicate the output where profits will be maximized. Sketch in the total profit curve.
   (b) In the lower part of the chart, using the same *x*-axis scale and any convenient scale on the *y*-axis, sketch in the corresponding average revenue, marginal revenue, average total cost, and marginal cost curves. Shade the area representing total profit at the most profitable output.
4. Reproduce the chart of the preceding question on the assumption that the firm sells in a less than perfectly competitive (monopolistic) market.
5. Draw on the same chart a total cost and a total variable cost curve. Locate the point on the *VC* curve where average variable cost is at a minimum, and call the point *P*. Locate the point on the *TC* curve where average total cost is at a minimum, and call the point *R*. Prove that *R* is to the right of *P*.

6. Draw a total cost curve and a total variable cost curve on the assumption that marginal cost is constant through a wide range.

7. Draw total cost, total variable cost, and total fixed cost curves on one chart. On another chart draw average total, average variable, average fixed, and marginal cost curves. Assume that both charts are drawn from the same basic data. Locate the following points on both charts:

(a) *AVC* when *AVC* is a minimum;
(b) *ATC* when *ATC* is a minimum;
(c) total fixed cost;
(d) output when *MC* is a minimum;
(e) output when *AVC* is a minimum;
(f) output when *ATC* is a minimum;
(g) *TC* when *MC* is a minimum;
(h) *VC* when *AVC* is a minimum;
(i) *TC* when *ATC* is a minimum;
(j) *AFC* when *ATC* is a minimum.

8. In Figure 15-3, *TC* is total cost, *VC* is variable cost, and *FC* is total fixed cost. $MC = \tan a = K$. But *ATC* at $x = Ox$ is $\tan b$, which is greater than *ATC* at $x = Ox'$, which in turn is greater than *ATC* at $x = Ox''$. Query: if *MC* is constant, why does not $MC = ATC$?

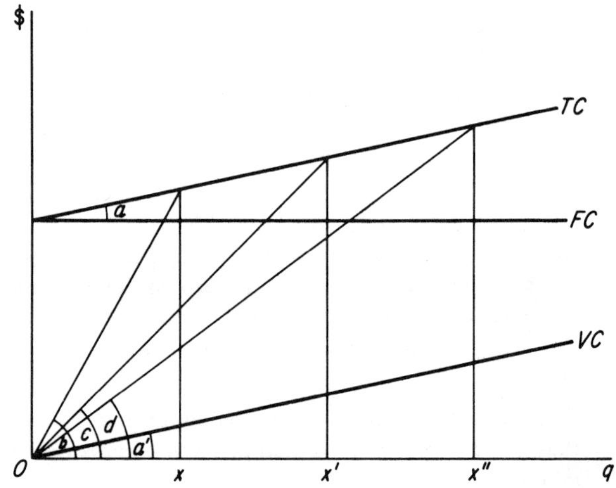

Fig. 15-3

9. Assume that *ATC* is constant at 5 from $x=4$ to $x=8$ and that total fixed cost $= 12$. Plot (a) *ATC*, (b) *AFC*, and (c) *AVC* from $x=4$ to $x=8$. On a second chart plot (d) *TC*, (e) *VC*, and (f) *FC*.

15.4 Bibliographical Note

In connection with this chapter the student should read the excellent discussion of total–average–marginal relationships in George J. Stigler, *The Theory of Price*, rev. ed., pp. 20–30. Also see Clark Lee Allen, Aurelius Morgner, and Robert H. Strotz, *Problem in the Theory of Price*, pp. 154–168. Total–average and total–marginal relationships become important in discussions of the law of variable proportions (diminishing returns). See, for example, Frank H. Knight, *Risk, Uncertainty, and Profit*, especially Chapter 4; John F. Due, *Intermediate Economic Analysis*, 3rd ed., pp. 123–134; Richard H. Leftwich, *The Price System and Resource Allocation*, rev. ed., Chapter 7; Sidney Weintraub, *Price Theory*, pp. 72–82.

16

Geometry of Point Elasticity of Demand

16.1 Point Elasticity

Elasticity of demand with respect to price is defined as $\Delta x/x \div \Delta y/y$. In Figure 16-1, elasticity at point R is

$$
\begin{aligned}
E_R &= \frac{MM'}{OM} \div \frac{RS}{MR} = \frac{MM'}{OM}\cdot\frac{MR}{RS} = \frac{SV}{OM}\cdot\frac{MR}{RS} = \frac{SV}{RS}\cdot\frac{MR}{OM} \\
&= \frac{MT'}{MR}\cdot\frac{MR}{OM} \\
&= \frac{MT'}{OM} \\
&= \frac{RT'}{RT} \\
&= \frac{OP}{TP}.
\end{aligned}
$$

16.2 Exercises

1. On Figure 16-2 find the point P' on D' which has the same elasticity as point P on curve D.

2. On Figure 16-3, compare the elasticity of AB and CD:
   (a) at price OP;
   (b) at quantity OM.

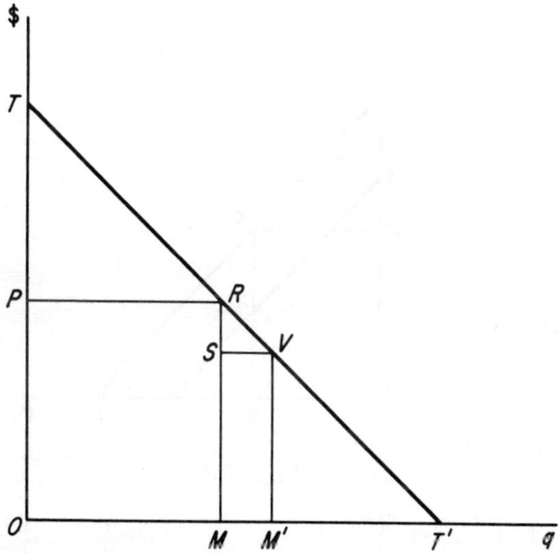

Fig. 16-1

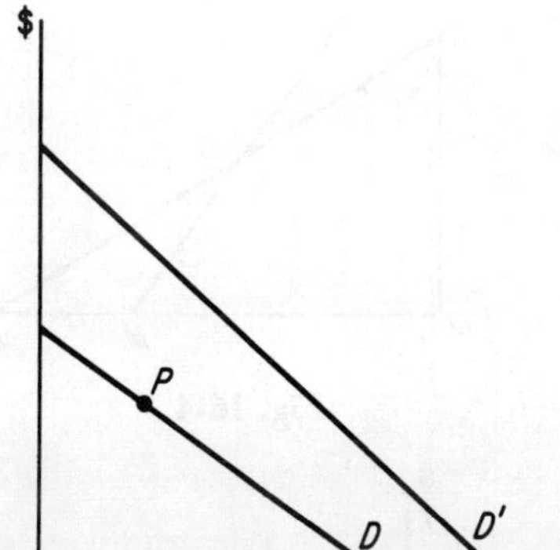

Fig. 16-2

3. The equation of a demand curve is $y=K/x$, where K is a constant. If a tangent to the demand curve crosses the y-axis at 12, at what price does the tangent touch the demand curve?

4. Demonstrate geometrically that $m=p(1+1/e)$, where m is marginal revenue, p is price, and e is elasticity.

5. On Figure 16-4, find the point D' on curve LM which has the same elasticity as point D on curve AB.

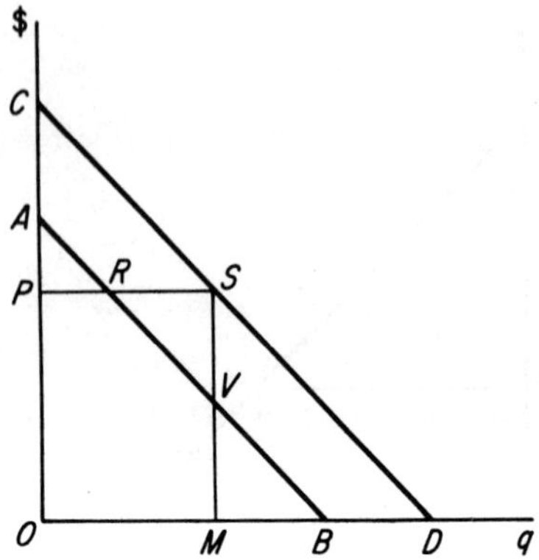

Fig. 16-3

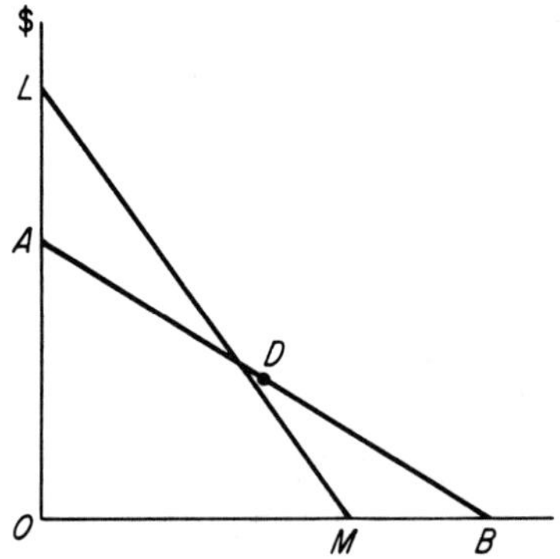

Fig. 16-4

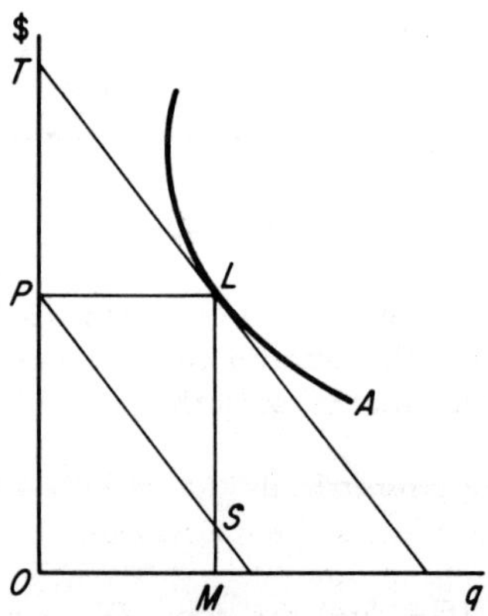

Fig. 16-5

6. Demonstrate geometrically that $e=a/(m-a)$, where e is elasticity of demand, a is average revenue, and m is marginal revenue. Also solve for a and m.

7. In both Figures 16-5 and 16-6, PS is parallel to TL and A is an average curve. (a) Prove in both cases that if average is equal to ML, marginal is equal to MS. (b) State generally how to determine geometrically the marginal value corresponding to a given average value.

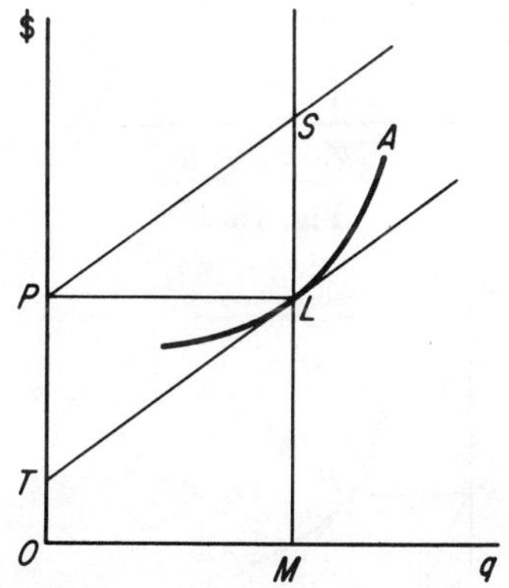

Fig. 16-6

8. If a linear demand curve has an x-intercept at T' and a y-intercept at T, show that the elasticity at any price, OP, where the quantity demanded is OM, is equal to

$$e = \frac{OT'}{OT} \cdot \frac{OP}{OM}$$

9. In Figure 16-7, show that the elasticity of demand at price OP is given by

$$e = \frac{\text{slope of } PM}{\text{slope of } TT'}.$$

10. In Figure 16-8, AB is parallel to RM. Prove that the elasticity of demand at price OR is equal to OC/OB.

11. Prove that if two linear demand curves are parallel, a line drawn from the origin through the two demand curves cuts them at points of equal elasticity.

12. Given: $qp^2=5040$, where $p=$price and $q=$quantity demanded.

    (a) Plot the demand curve.

    (b) What is the elasticity of demand at any price?

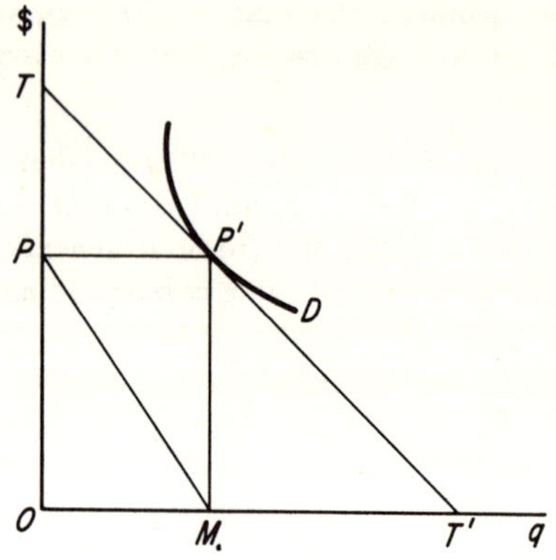

Fig. 16-7

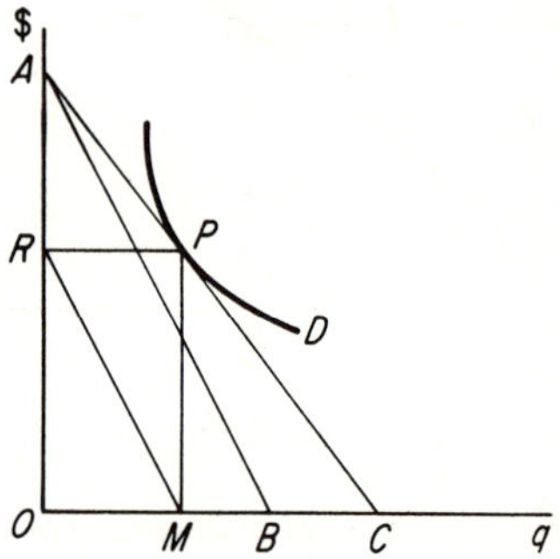

Fig. 16-8

13. The equation for a straight-line demand curve is $Q = a(b - P)$, where Q = quantity demanded, P = price, and a and b are constants.

(a) What is the intercept of the demand curve on the y-axis?

(b) What is the intercept of the demand curve on the x-axis?

(c) Show that the elasticity of demand at any price is equal to

$$e = \frac{P}{P-b}.$$

(d) What is the elasticity when $P = b/2$?

(e) Show that

$$P = \frac{eb}{e-1}.$$

(f) At what price will $e = -3$?

14. Given:

$$p = \frac{eb}{e-1} \quad \text{and} \quad p = m\left(\frac{e}{e+1}\right),$$

where p = price, e = elasticity, m = marginal revenue, and b = the y-intercept of the demand curve.

(a) Show that $e = (m+b)/(m-b)$.

(b) Show that for a perfectly competitive firm $e = -\infty$.

(c) Solve for m in terms of b and e.

(d) Solve for b in terms of m and e.

15. Given: $e = P/(P-b)$ and $e = a/(m-a)$, where e = elasticity, b = the y-intercept of a linear demand curve, P = price, a = average revenue, and m = marginal revenue.

(a) Prove that $P = (b+m)/2$.

(b) Assume that the price changes from P to P' and marginal revenue changes from m to m'. Show that $\Delta P = \Delta m/2$.

16. If the x-intercept of a linear demand curve is a, show that the elasticity of demand at quantity $= x$ is equal to $e = 1 - a/x$.

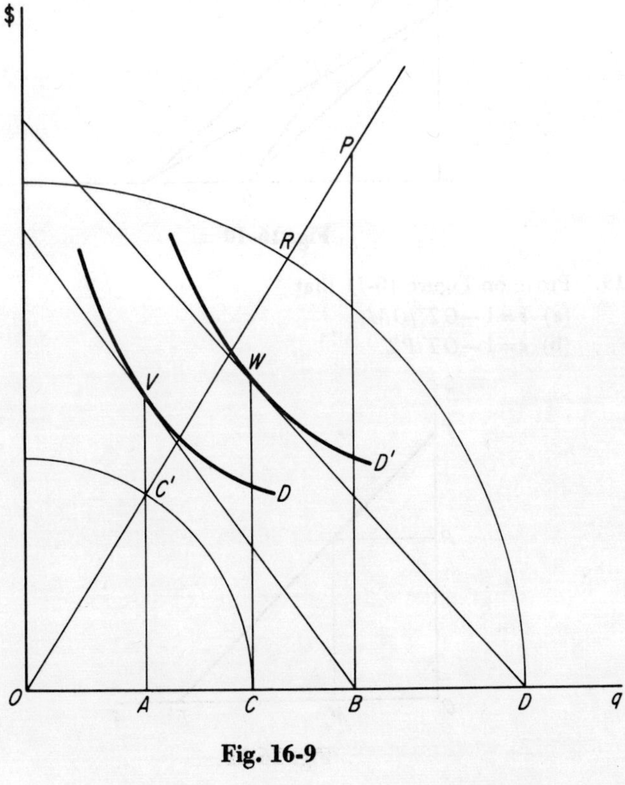

Fig. 16-9

17. The problem is to compare the elasticity of D at point V with the elasticity of D' at point W in Figure 16-9. Confirm the following:

$$E_V = \frac{AB}{OA} = \frac{C'P}{OC'}$$

$$E_W = \frac{CD}{OC} = \frac{C'R}{OC'}$$

$$E_V : E_W = \frac{C'P}{OC'} : \frac{C'R}{OC'} = C'P : C'R$$

18. (a) Find the points on curves AB, DB, and DC where the elasticity is the same as at point P on curve AC of Figure 16-10.

(b) Find the points on curves DC, AC, and AB where the elasticity is the same as at point P on curve DB.

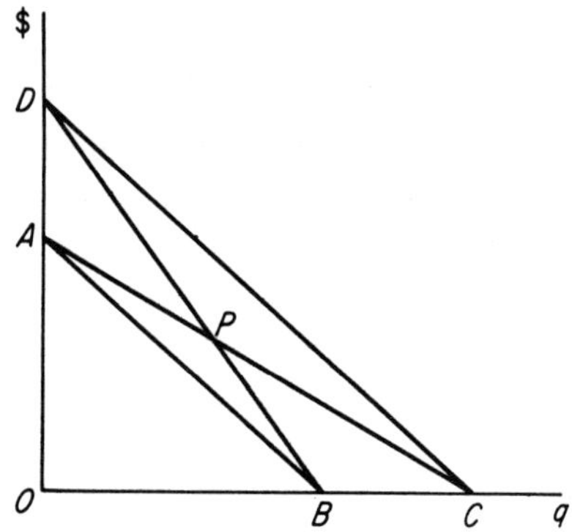

Fig. 16-10

19. Prove on Figure 16-11 that

(a) $e = 1 - OT'/OM$;

(b) $e = 1 - OT/PT$.

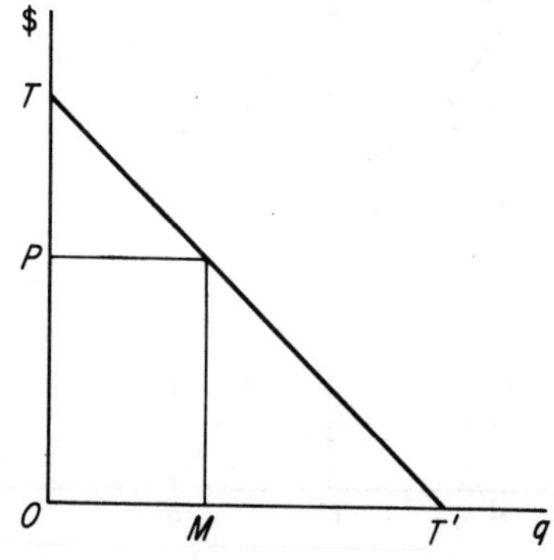

Fig. 16-11

16.3 Bibliographical Note

Every student of price theory should be familiar with the analysis of elasticity developed in Alfred Marshall, *Principles of Economics*, 8th ed., Mathematical Appendix, Note 3. Most elementary and intermediate economic theory texts develop the idea of elasticity geometrically; the following may be consulted: Clark Lee Allen, James M. Buchanan, and Marshall R. Colberg, *Prices, Income, and Public Policy*, 2nd ed., pp. 479–484; Joe S. Bain, *Price Theory*, pp. 40–50; John F. Due, *Intermediate Economic Analysis*, 3rd ed., pp. 89–95; Richard H. Leftwich, *The Price System and Resource Allocation*, rev. ed., pp. 34–45; George J. Stigler, *The Theory of Price*, rev. ed., pp. 30–38; Sidney Weintraub, *Price Theory*, pp. 42–46.

17

Additional Applications of the Elasticity Concept

17.1 Arc Elasticity

Point elasticity, as we have seen, is quite specific and measurable. But frequently the economist and often the businessman is less interested in elasticity at a point than in elasticity between two points. For this purpose the concept of arc elasticity has been devised. Unfortunately, arc elasticity is less precise than point elasticity; at best it represents a kind of average of elasticities, and the specific value one gets depends upon which of several formulas one uses. Among the arc-elasticity formulas that have been devised are the following:

$$\frac{p}{q}\cdot\frac{\Delta q}{\Delta p} \tag{1}$$

$$\frac{p+\Delta p}{q}\cdot\frac{\Delta q}{\Delta p} \tag{2}$$

$$\frac{p}{q+\Delta q}\cdot\frac{\Delta q}{\Delta p} \tag{3}$$

$$\frac{p+\Delta p}{q+\Delta q}\cdot\frac{\Delta q}{\Delta p} \tag{4}$$

$$\frac{q_1-q_2}{q_1+q_2}\cdot\frac{p_1+p_2}{p_1-p_2} \tag{5}$$

Of these the last is probably the most commonly employed and is derived as follows:

$$e = \frac{q_1-q_2}{\frac{q_1+q_2}{2}} \div \frac{p_1-p_2}{\frac{p_1+p_2}{2}}$$

$$= \frac{q_1-q_2}{q_1+q_2} \div \frac{p_1-p_2}{p_1+p_2}$$

$$= \frac{q_1-q_2}{q_1+q_2} \cdot \frac{p_1+p_2}{p_1-p_2}.$$

17.2 Exercises

1. Consider the demand function: $5p+q=100$.
   (a) Compute arc elasticity between $p=12$ and $p=10$ by each of the formulas above. (In formulas (1) to (4) let $p=12$.)
   (b) Determine point elasticity at $p=12$, $p=10$, and $p=11$.
   (c) Show that formula (5) above actually gives point elasticity at the mid-point of a straight line drawn from one point on the curve to the other.

2. Suppose that at point P on a demand curve $p=50$ and $q=100$, and at point P' on the same demand curve $p=50$ and $q=150$. What is the arc elasticity between points P and P'?

3. Suppose that at point R on a demand curve $p=10$ and $q=50$, and at point R' on the same demand curve $p=5$ and $q=50$. What is the arc elasticity between R and R'?

4. Given: $pq=1200$. Determine arc elasticity of demand between $p=12$ and $p=10$ by each of the arc elasticity formulas. (In formulas (1) to (4) let $p=12$.)

5. Plot the demand function $y=12/x$ from $x=1$ to $x=6$. On the same chart plot the linear demand function $y=10-2x$. Determine the arc elasticity of each curve from $y=6$ to $y=4$, using formula (5) above.

6. Plot the demand function $y=(6x+12)/x$ from $x=1$ to $x=6$. On the same chart plot the linear demand function $y=16-2x$. Determine the arc elasticity of each curve from $x=2$ to $x=3$, using formula (5) above.

17.3 Cross Elasticity

The concept of cross elasticity is useful as a measure of the relationship between the demands for two or more commodities. The elasticity of the demand for X in terms of the price of Y is defined as

$$E_{xpy} = \frac{\text{relative change in quantity of } X}{\text{relative change in price of } Y},$$

the price of X being held constant. When commodities are substitutes for each other, their cross elasticities are positive. If, for example, the price of butter increases, the consumption of margarine will also increase. When commodities are complementary to each other, their cross elasticities are negative. If the price of gasoline decreases, the consumption of tires increases.

The following is a formula for measuring the cross elasticity of X in terms of the price of Y:

$$E_{xpy} = \frac{q_1 - q_2}{q_1 + q_2} \cdot \frac{p_1 + p_2}{p_1 - p_2},$$

where the q's are the quantities of X and the p's are the prices of Y.

17.4 Income Elasticity

In addition to observing the price elasticity of demand, we may sometimes have occasion to note the income elasticity of demand, which is defined as

$$E_{qm} = \frac{\text{relative change in quantity}}{\text{relative change in income}}.$$

In general, as income increases the consumption of most commodities increases. That is to say that the quantity demanded at a given price will increase as income increases. For some goods, however, the amount consumed at high incomes is less than the amount consumed at lower incomes, the price remaining the same. Examples of such "inferior" goods are lard and the cheaper cuts of meat.

17.5 Elasticity of Total Cost

The elasticity concept has numerous applications in addition to its use as a measure of the responsiveness of demand or supply to changes in price or income. The elasticity of total cost, for example, may be defined as the responsiveness of total cost to changes in quantity:

$$E_{tc} = \frac{\Delta TC}{TC} \div \frac{\Delta X}{X} = \frac{\Delta TC}{TC} \cdot \frac{X}{\Delta X} = \frac{\Delta TC}{\Delta X} \cdot \frac{X}{TC} = \frac{\Delta TC}{\Delta X} \div \frac{TC}{X}.$$

But since $\Delta TC/\Delta X$=marginal cost and TC/X=average cost, we may write

$$E_{tc} = \frac{MC}{AC}.$$

The elasticity of total cost may also be determined geometrically. In Figure 17-1, TC is a total cost curve. The elasticity of total cost from $x = OM$ to $x = ON$ may be determined as follows:

$$E_{tc} = \frac{\Delta TC}{TC} \div \frac{\Delta X}{X} = \frac{SR}{ML} \div \frac{LS}{OM} = \frac{SR}{ML} \cdot \frac{OM}{LS} = \frac{SR}{LS} \cdot \frac{OM}{ML} = \frac{ML}{AM} \cdot \frac{OM}{ML}$$

$$= \frac{OM}{AM} = \frac{TL}{AL} = \frac{TP}{OP}.$$

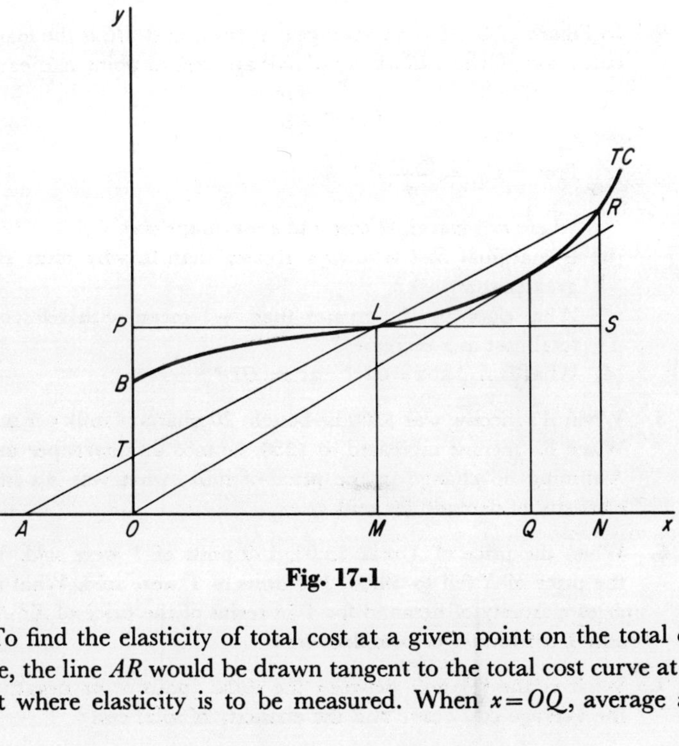

Fig. 17-1

To find the elasticity of total cost at a given point on the total cost curve, the line AR would be drawn tangent to the total cost curve at the point where elasticity is to be measured. When $x = OQ$, average and marginal are equal, and the elasticity of total cost is equal to 1; for values of x less than OQ, E_{tc} is less than 1; and for values of x greater than OQ, E_{tc} is greater than 1.

17.6 Exercises

1. Demonstrate algebraically that the elasticity of total product is equal to

$$E_{tp} = \frac{MP}{AP},$$

where MP = marginal product and AP = average product.

2. Show geometrically that the elasticity of total product is equal to

$$E_{tp} = \frac{TP}{OP},$$

when OP = total product at a given input and T is the y-intercept of the tangent to the total product curve at the given input.

3. Show that for a firm in a perfectly competitive industry the elasticity of total revenue is equal to 1 for all values of x.

4. In Figure 17-2, AC is an average cost curve and MC is the marginal cost curve. Given: Elasticity of average cost at point R is equal to

$$E_{ac} = \frac{TP}{OP}.$$

(a) Prove: $E_{ac} = \frac{m-a}{a}$,

where m = marginal cost and a = average cost.

(b) If marginal cost is always greater than 0, why must E_{ac} be greater than -1?

(c) What does an E_{ac} greater than -1 mean with reference to total cost as x increases?

(d) What is E_{ac} at $x = OV$? at $x = OL$?

5. When A's income was \$300 he bought 20 quarts of milk per month. When his income increased to \$350, he took 24 quarts per month. Assuming no change in the price of milk, what was A's income elasticity of demand for milk?

6. When the price of X was \$5.00, 100 units of Y were sold. When the price of X fell to \$4.00, 120 units of Y were sold. What is the cross elasticity of demand for Y in terms of the price of X? Are X and Y substitutes or complements?

7. What is the relation between the slope (positive or negative) of the average cost curve and the elasticity of total cost?

8. Show that if the demand curve is linear and negatively inclined the elasticity of total revenue is always less than 1, but to the left of its maximum value the elasticity of total revenue is positive and to the right of the maximum value the elasticity of total revenue is negative.

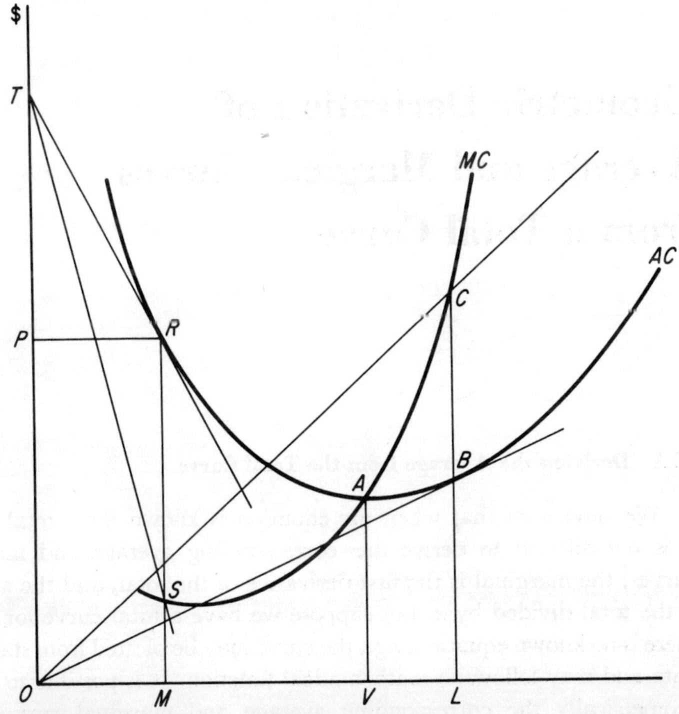

Fig. 17-2

17.7 Bibliographical Note

The concept of arc elasticity is discussed in Joe S. Bain, *Price Theory*, pp. 42–43; Ralph K. Davidson, Vernon L. Smith, and Jay W. Wiley, *Economics: an Analytical Approach*, pp. 20–21; John F. Due, *Intermediate Economic Analysis*, 3rd ed., pp. 89–95; Richard H. Leftwich, *The Price System and Resource Allocation*, rev. ed., pp. 35–38; George J. Stigler, *The Theory of Price*, rev. ed., pp. 35–36; Sidney Weintraub, *Price Theory*, pp. 46–48. Cross elasticity is discussed in Bain, pp. 50–53; Due, pp. 99–101; Leftwich, pp. 45–46; Stigler, pp. 48–49; Alfred W. Stonier and Douglas C. Hague, *A Textbook of Economic Theory*, pp. 85–86; Weintraub, 48–49. The elasticity of cost is discussed in R. G. D. Allen, *Mathematical Analysis for Economists*, pp. 260–264. Income elasticity is discussed in Bain, pp. 54–56; Due, pp. 98–99; Stigler, pp. 49–53; Stonier and Hague, pp. 70—72.

18

Geometric Derivation of Average and Marginal Curves from a Total Curve

18.1 Deriving the Average from the Total Curve

We have seen that when the equation is known for a total curve, it is not difficult to derive the corresponding average and marginal curves; the marginal is the first derivative of the total, and the average is the total divided by x. But suppose we have a total curve for which there is no known equation, *e.g.*, the curve may be plotted from statistical data and may follow no mathematical function. It is possible to derive geometrically the corresponding average and marginal curves, and doing so provides a good exercise in average-marginal-total relationships.

In Figure 18-1, the total curve is the function $y=f(x)=12x-2x^2$. A curve with a known function is used so that we may check the accuracy of our method since we know that the corresponding average must be $12-2x$ and the marginal must be $y=12-4x$.

To determine the average curve corresponding to a given total curve, follow these directions: First, mark off vectors from the origin, passing through the total curve. The vectors are drawn so that for any given x value the y value of the second vector is twice that of the first, the y value of the third is three times that of the first, and so on. In the figure these vectors have been numbered from 1 to 10.

At the point on the total curve where vector 1 cuts it (where $x=5\frac{1}{2}$ in the figure) the average value is the tangent of the angle formed by vector 1 and the x-axis. We may arbitrarily call this average $=1$. Where vector 2 cuts the total curve (at $x=5$), the average value is the tangent of the angle formed by vector 2 and the x-axis. At this x-value, the

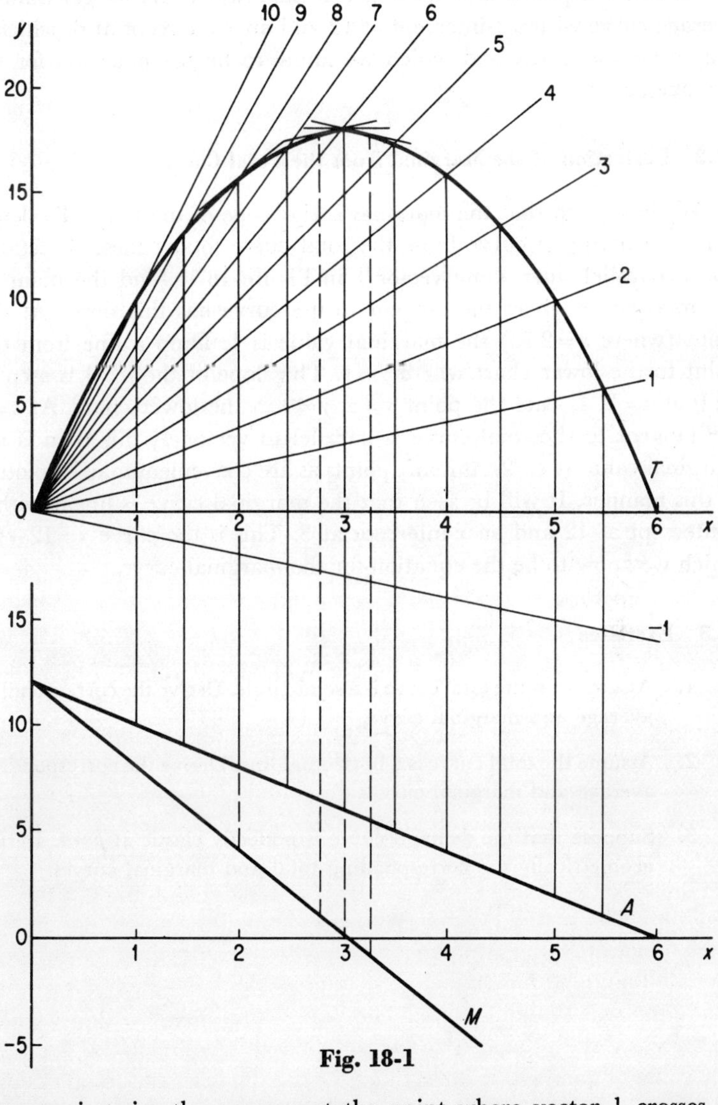

Fig. 18-1

average is twice the average at the point where vector 1 crosses the total curve. Since the first average was given the value 1, average at $x=5$ must be 2, and so on.

From the point where vector 1 cuts the total curve, drop a line to the lower chart where $y=1$. From the point where vector 2 cuts the total curve, drop a line to the lower chart where $y=2$, and so on. By connecting the points so determined in the lower chart we get a linear average curve with a y-intercept at 12 and an x-intercept at 6, which is the curve $y=12-2x$ and which we know to be the equation for the average curve.

18.2 Derivation of the Marginal from the Total Curve

We have seen that the marginal is the slope of the total. To determine the marginal curve from the total curve follow these directions. Lay a parallel ruler along vector 1 in Figure 18-1. Find the point on the total curve where the tangent to the curve has this slope. At this point (where $x=2.75$) the marginal value is 1. Drop a line from this point to the lower chart where $y=1$. The slope of the total is seen to be 0 at $x=3$. Locate the point $x=3$, $y=0$ in the lower chart. At $x=2$ the tangent to the total curve is parallel to vector 4; this then is the marginal value at $x=2$. As many points as are convenient may be plotted in this manner. It will be seen that the marginal curve is linear with a y-intercept at 12 and an x-intercept at 3. This is the curve $y=12-4x$, which we know to be the equation for the marginal curve.

18.3 Exercises

1. Assume that the total curve is a semi-circle. Derive the corresponding average and marginal curve.
2. Assume the total curve is a horizontal line. Derive the corresponding average and marginal curves.
3. Suppose that the demand curve is perfectly elastic at $p=4$. Derive geometrically the corresponding total and marginal curves.

19

The Nature of Indifference Curves

19.1 Characteristics of Indifference Systems

A relatively new tool that has proved useful in price theory is indifference curves. Its advantage lies in the fact that it is scientifically simpler—that is, it makes fewer assumptions—than alternate devices. The concept of marginal utility, for example, implies an ability to measure utility cardinally; if marginal utility is 6, this means that the consumption of the marginal unit has added 6 utility units to the consumer's satisfaction. But there is no way to measure units of satisfaction; the marginal utility concept purports to give a measure of what in fact cannot be measured. The indifference-curve approach to the problem of consumer satisfaction avoids the necessity of making cardinal measures of utility. What is possible, it is assumed, is for a consumer to determine that one combination of goods is preferable to another or that both combinations are equally attractive. If one combination is better than another, the indifference-curve technique does not require a determination of how much better one is than the other. That is to say, ordinal measures of utility are possible even though cardinal measures are not. The use of indifference curves requires only ordinal measures. Any point on a given indifference curve represents a combination of commodities X and Y that will provide an equal degree of satisfaction or utility to the consumer. Any point on a "higher" indifference curve will provide more satisfaction to the consumer than any point on a "lower" indifference curve. And this is all that we need to know; we do not have to determine how much better a point on one indifference curve is than a point on another curve.

The principal characteristics of indifference systems are the following: 1. Indifference curves slope downward to the right. 2. Indifference curves are convex to the origin. 3. Indifference curves cannot intersect. 4. Any point in an indifference field lies on one and only one indifference curve. Each of these points will be discussed briefly.

1. The negative slope of an indifference curve implies that by getting more of commodity X and giving up an appropriate amount of Y the consumer is as well off as he was originally. If the indifference curve had a positive slope, it would imply that greater amounts of both X and Y would leave the consumer no better off than before. A horizontal indifference curve would indicate that the consumer would have no preference between a given amount of Y and varying amounts of X. And a vertical indifference curve would indicate that the consumer had no preference between a given amount of X and varying quantities of Y.

2. The convexity of an indifference curve reflects the decreasing marginal rate of substitution of X for Y (written MRS_{xy}). If we start with combination R_1 in Figure 19-1, which comprises a relatively small

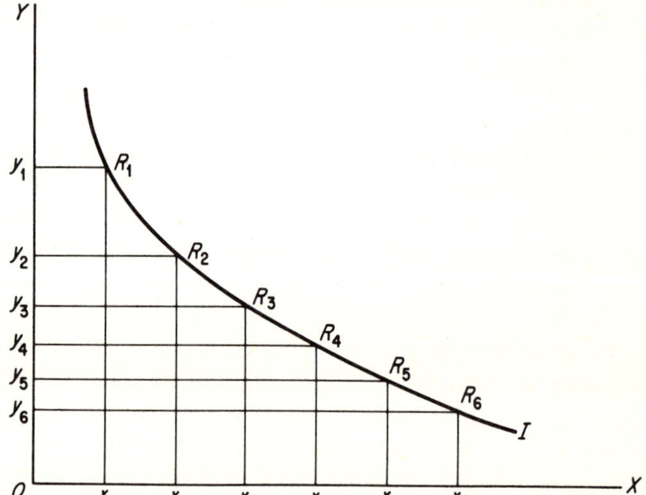

Fig. 19-1

quantity of X and a large quantity of Y, the assumption is that the consumer could give up a considerable quantity of Y in order to obtain an additional unit of X and be just as well off as he was before. But as he acquires more and more of X and has less and less of Y, the amount of Y he will be willing to give up in order to get one more unit of X becomes increasingly smaller. The *MRS* is defined as the amount of Y the consumer is willing to sacrifice to obtain one more unit of X.

3. If two indifference curves intersected, it would imply that certain combinations of X and Y were equal to other combinations, some of which represented higher and some lower preferences. Since one combination cannot at the same time be equal to another and also be better or worse than the other, intersecting indifference curves would represent pure nonsense.

4. Indifference curves cannot intersect, but it is not necessary that they be parallel; two indifference curves may be farther apart at some points than at others. But if every point in an indifference field lies on

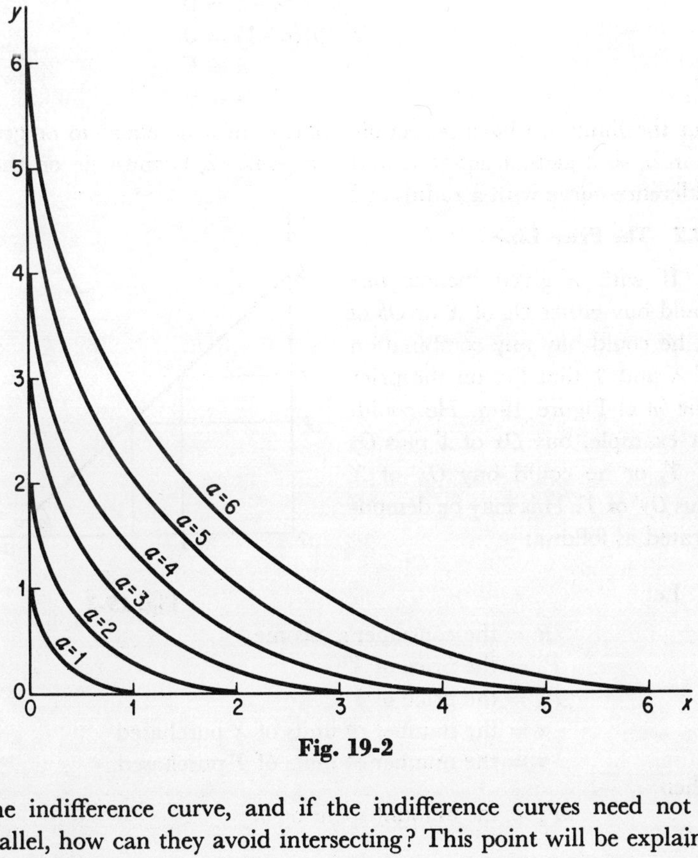

Fig. 19-2

some indifference curve, and if the indifference curves need not be parallel, how can they avoid intersecting? This point will be explained in some detail.

Suppose that the following is the equation for a family of indifference curves:

$$(x-a)^2+(y-a)^2 = a^2,$$

where a is a parameter, and $0 \leqslant x \leqslant a$. This is the equation for a circle with a radius $= a$, the center having an x-value $= a$ and a y-value $= a$. In Figure 19-2, curves are drawn with a ranging from 1 to 6.

It is clear that these indifference curves are not parallel; each is the arc of a circle, and each circle has a different radius. Within the limits set, *i.e.*, $0 \leqslant x \leqslant a$, the curves cannot intersect. And any point lies on some indifference curve. Consider, for example, the point (2, 1). We may substitute $x=2$ and $y=1$ in the equation $(x-a)^2+(y-a)^2=a^2$:

$$(2-a)^2+(1-a)^2 = a^2$$
$$4-4a+a^2+1-2a+a^2 = a^2$$
$$a^2-6a+5 = 0$$
$$(a-5)(a-1) = 0$$
$$a = 1$$

or

$$a = 5.$$

But the limits we have set require that a must be equal to or greater than x, so a cannot equal 1, and the point (2, 1) must lie on the indifference curve with a radius of 5.

19.2 The Price Line

If with a given income one could buy either Oa of X or Ob of Y, he could buy any combination of X and Y that lies on the price line ba of Figure 19-3. He could, for example, buy Ox of X plus Oy of Y, or he could buy Ox' of X plus Oy' of Y. This may be demonstrated as follows:

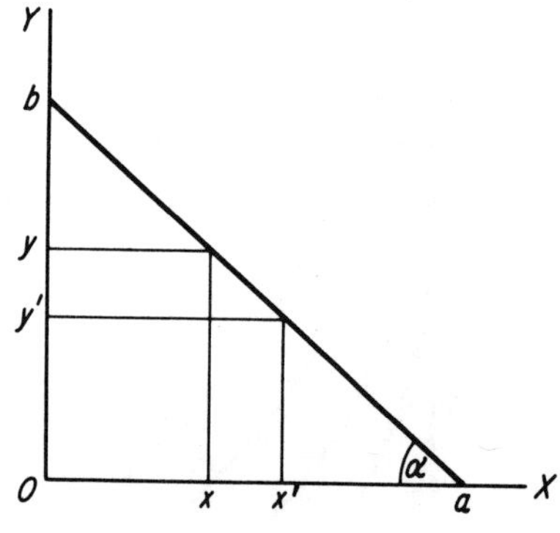

Fig. 19-3

Let

$$R = \text{the consumer's income}$$
$$P_x = \text{the price of } X$$
$$P_y = \text{the price of } Y$$
$$x = \text{the number of units of } X \text{ purchased}$$
$$y = \text{the number of units of } Y \text{ purchased.}$$

Then

$$xP_x = \text{the amount spent on } X$$

and

$$yP_y = \text{the amount spent on } Y.$$

If the entire income, R, is spent on X and Y,

$$R = xP_x + yP_y$$

$$1 = \frac{xP_x}{R} + \frac{yP_y}{R}$$

$$1 = \frac{x}{R/P_x} + \frac{y}{R/P_y}.$$

Since

$$R/P_x = a \quad \text{and} \quad R/P_y = b,$$

$$1 = \frac{x}{a} + \frac{y}{b},$$

which is the equation for a straight line with an intercept on the x-axis at a and an intercept on the y-axis at b, *e.g.*, ba in Figure 19-3. The price line is sometimes called the budget line, the opportunity line, or the line of attainable combinations.

The slope of the price line is minus the price of X in terms of Y. If the consumer can with the same total expenditure buy either Oa of X or Ob of Y, the price of X in terms of Y is

$$\text{slope of } ba = -\tan\alpha = \frac{-Ob}{Oa} = \frac{-R/P_y}{R/P_x} = \frac{-R}{P_y}\cdot\frac{P_x}{R} = \frac{-P_x}{P_y}.$$

If the price of X should fall, a would move to the right, and the slope of ba would be increased, *i.e.*, a reduced negative value.

19.3 Indifference Curves and Elasticity of Demand

There are many applications of indifference curves in price theory and welfare economics. One interesting use of this technique that has been largely ignored by economists is in the measurement of the elasticity of demand. Consider the indifference curves in Figure 19-4. If, as the price of X changes from that indicated by price line AB to that indicated by price line AB', the consumer moves from point R on indifference curve I to point S on indifference curve II, we know that the elasticity of demand between these two prices is equal to one since the total expenditure on X remains constant at TA as the price changes. This may be demonstrated somewhat more rigorously.

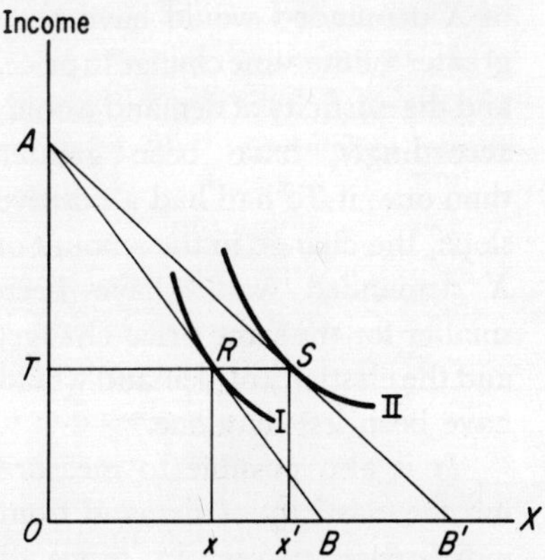

Fig. 19-4

Let us begin with the formula for arc elasticity:

$$E = \frac{q_1 - q_2}{q_1 + q_2} \cdot \frac{p_1 + p_2}{p_1 - p_2}.$$

Then in terms of Figure 19-4

$$E = \frac{Ox - Ox'}{Ox + Ox'} \cdot \frac{\frac{OA}{OB} + \frac{OA}{OB'}}{\frac{OA}{OB} - \frac{OA}{OB'}}$$

$$= \frac{-xx'}{Ox + Ox'} \cdot \frac{\frac{OA \cdot OB' + OA \cdot OB}{OB \cdot OB'}}{\frac{OA \cdot OB' - OA \cdot OB}{OB \cdot OB'}}$$

$$= \frac{-xx'}{Ox + Ox'} \cdot \frac{OB' + OB}{OB' - OB} = \frac{-xx'}{Ox + Ox'} \cdot \frac{OB' + OB}{BB'}.$$

But

$$\frac{-xx}{Ox + Ox'} = \frac{-RS}{TR + TS} = \frac{-BB'}{OB + OB'}.$$

Therefore,

$$E = \frac{-BB'}{OB + OB'} \cdot \frac{OB' + OB}{BB'} = -1.$$

We may generalize by saying that if the line *TS* had had a negative slope, the change in the amount of *X* demanded would have been greater for the same change in price, and the elasticity of demand would, accordingly, have been greater than one; if *TS* had had a positive slope, the change in the amount of *X* demanded would have been smaller for the same price change, and the elasticity of demand would have been less than one.

It is also possible to measure income elasticity of demand from indifference curves. In terms of Figure 19-5,

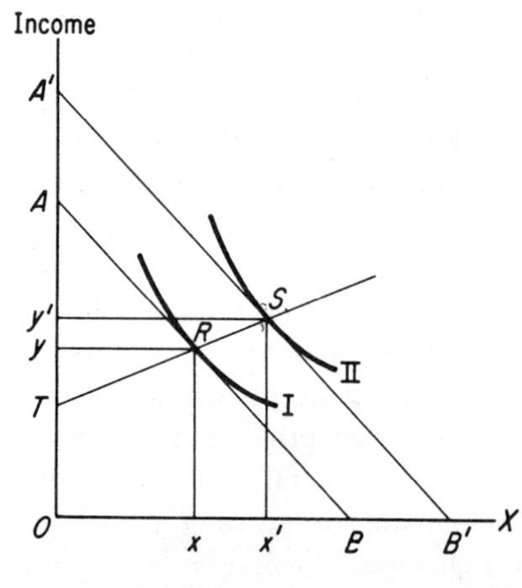

Fig. 19-5

$$E_y = \frac{\Delta x}{x} \div \frac{\Delta y}{y}$$
$$= \frac{xx'}{Ox} \div \frac{AA'}{OA} = \frac{xx'}{Ox} \cdot \frac{OA}{AA'}.$$

But

$$\frac{xx'}{Ox} = \frac{RS}{TR} = \frac{AA'}{TA}.$$

Therefore,

$$E_y = \frac{AA'}{TA} \cdot \frac{OA}{AA'} = \frac{OA}{TA} > 1.$$

We may generalize by saying that if OA is the original income and T is the y-intercept of the linear income-consumption curve, the income elasticity of demand may be measured by the formula

$$E_y = \frac{OA}{TA}.$$

19.4 Exercises

1. Draw two intersecting indifference curves and demonstrate the logical inconsistency involved.
2. Show that the point of tangency between the price line and an indifference curve determines the optimum combination of X and Y for the consumer.
3. Suppose that the amount of commodity X is measured on the x-axis and the consumer's income is measured on the y-axis. Draw the price line and indifference curve to indicate that the consumer is maximizing his satisfaction. Indicate on the figure the consumer's total expenditure on commodity X.
4. Show that, if indifference curves were concave to the origin, the point of tangency between the price line and an indifference curve would establish a point of minimum well-being.
5. Demonstrate that when the consumer has maximized his welfare with income and prices given, the price of X in terms of Y is equal to minus the marginal rate of substitution of X for Y.
6. (a) Show that a consumer's welfare position is improved as a result of a decline in the price of X.
   (b) Indicate how much income would have to be increased in order to improve the consumer's welfare position equally if there were no change in prices.

7. Draw families of indifference curves indicating that X and Y are:
   (a) perfect substitutes;
   (b) perfect complements.

8. Draw a consumer's indifference map and show the price-consumption curve.

9. Draw a consumer's indifference map and show the income-consumption curve.

10. Show how a demand curve can be derived from indifference curves.

11. Draw a family of indifference curves with appropriate price lines to indicate that the demand for X has an elasticity of -1.

12. Using the price lines in Figure 19-6, draw in appropriate indifference curves to indicate two simultaneous changes:
   (a) the consumer's income increases;
   (b) the price of X decreases.

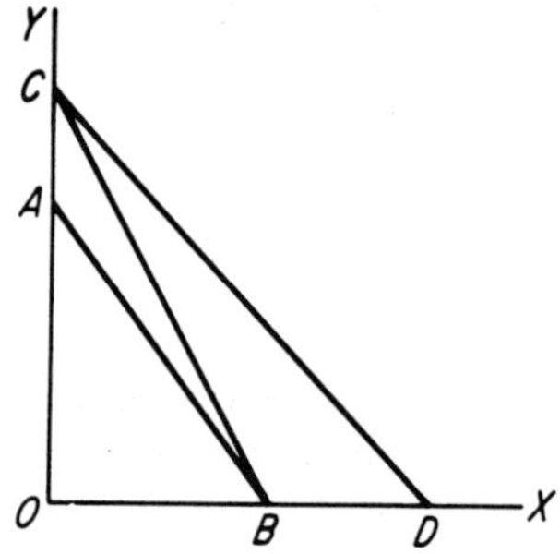

Fig. 19-6

13. In Figure 19-7, the consumer maximizes his satisfaction with combination L. Let income $= R$, $P_x = R/Oa$, and $P_y = R/Ob$.

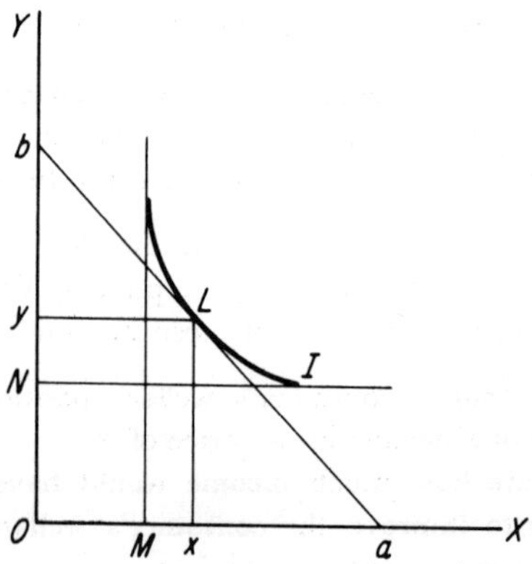

Fig. 19-7

(a) Suppose that P_x increases and P_y decreases. What is the highest price the consumer can pay for X without suffering a loss of well-being, assuming that P_y decreases appropriately? What then would be P_y?

(b) In terms of quantities of X and Y available to the consumer, what is necessary to be sure that his well-being has improved?

(c) Show that any price line other than ba passing through L would make possible an improvement in the consumer's well-being.

(d) If P_x increases so that if the entire income were spent on X, less than Ox of X could be bought, would the consumer's well-being necessarily be reduced?

(e) If P_y increases so that if the entire income were spent on Y, less than Oy of Y could be bought, would the consumer's well-being necessarily be reduced?

14. Draw the family of indifference curves with the following equations:

(a) $xy=10$;
(b) $xy=12$;
(c) $xy=15$.

Observe that (1) the indifference curves slope downward to the right; (2) the indifference curves are convex to the origin; and (3) the indifference curves cannot intersect.

(d) Show that the following points are on one and only one indifference curve: (3, 5); (3, 4); (2, 6); (2, $7\frac{1}{2}$); (4, $2\frac{1}{2}$).

15. Demonstrate that the formula for income elasticity of demand, $E_y=OA/TA$, is also valid:

(a) when T lies below the origin;
(b) when T is at the origin.

16. Draw parallel price lines. Label the one representing the lower income AA' and the other BB'. Draw indifference curve I tangent to AA' at L. Draw a vector from the origin through L and cutting BB' at M. Show that:

(a) if indifference curve II is tangent to price line BB' at M, income elasticity of demand equals one;
(b) if indifference curve II is tangent to BB' below M, income elasticity of demand is greater than one;
(c) if indifference curve II is tangent to BB' above M, income elasticity of demand is less than one.

19.5 Bibliographical Note

The student is referred to the classic works on indifference curves: J. R. Hicks and R. G. D. Allen, "A Reconsideration of the Theory of Value."

Economica, 1934, pp. 52–76 and 196–219, and J. R. Hicks, *Value and Capital*, 2nd ed., Chapters 1–3. If this is the student's first experience with indifference curves, he may want to consult one of the current intermediate textbooks on this subject. Among them are: Kenneth E. Boulding, *Economic Analysis*, 3rd ed., Chapters 36–37; John F. Due, *Intermediate Economic Analysis*, 3rd ed., pp. 77–89; Richard H. Leftwich, *The Price System and Resource Allocation*, rev. ed., pp. 67–90; Joseph P. McKenna, *Intermediate Economic Theory*, Chapter 7; Ruby Turner Norris, *The Theory of Consumer's Demand*, Chapter 2; George J. Stigler, *The Theory of Price*, rev. ed., Chapter 5; Alfred W. Stonier and Douglas C. Hague, *A Textbook of Economic Theory*, Chapter 3; Sidney Weintraub, *Price Theory*, Chapter 1.

20

Isoquants and the Production Function

20.1 The Nature of Isoquants

Isoquants (*iso* = equal; *quant* = quantity), sometimes called isoproduct curves, bear a family resemblance to indifference curves. An isoquant shows the various combinations of the factors X and Y that may be used to produce a given quantity of output. In Figure 20-1, a total product of 200 units per time period can be produced with Oy of Y plus Ox of X, or with Oy of Y plus Ox' of X, or with Oy' of Y plus Ox of X, or with any other combination of X and Y lying on the isoquant labeled 200. The figure also indicates that if the firm uses Oy of Y in combination with X, the greatest possible output is 300 units; this output will be reached when Oy of Y is combined with Ox'' of X; if either more or less than Ox'' of X is used with Oy of Y, the total product will be less than 300. It follows, accordingly, that even if X were available to the firm at no cost, it would never use more than Ox'' of X in combination with Oy of Y. It is also clear that the firm would never use Oy' of Y in combination with Ox of X, since the same output could be produced with Ox of X plus only Oy of Y; in fact, an output of 300 could be produced with less than Oy' of Y plus Ox of X.

20.2 Some Terms Defined

In Figure 20-2, the line OA is drawn through the several isoquants at the point where each isoquant would be tangent to a vertical line.

Since *OA* cuts the isoquants at points of equal slope, it is called an *isocline*. *OB* is also an isocline; it is drawn through the isoquants at the points where they would be tangent to horizontal lines. These particular

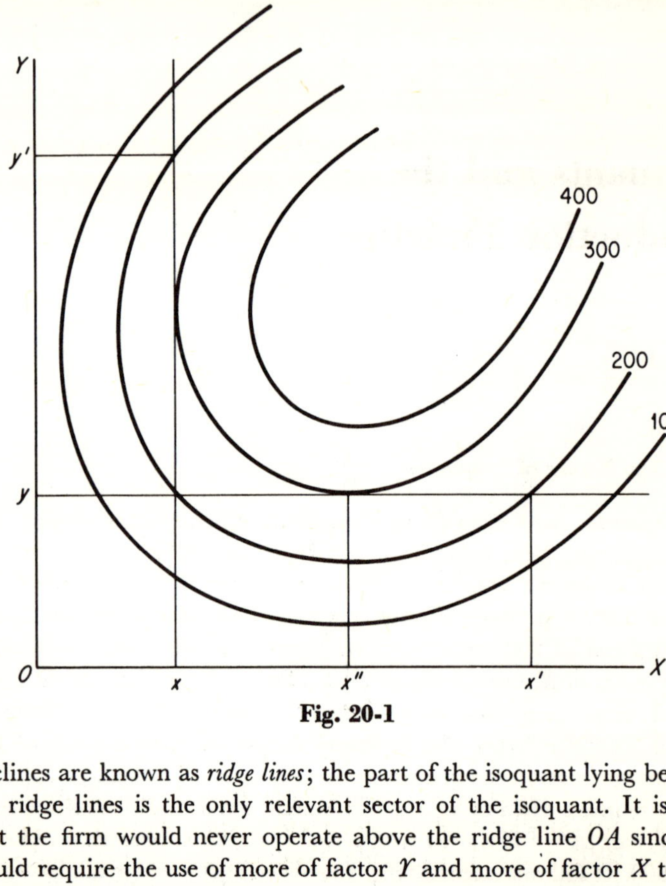

Fig. 20-1

isoclines are known as *ridge lines*; the part of the isoquant lying between the ridge lines is the only relevant sector of the isoquant. It is clear that the firm would never operate above the ridge line *OA* since this would require the use of more of factor *Y* and more of factor *X* to produce the same total output that could be produced with fewer factors. By the same token the firm would never operate to the right of ridge line *OB*.

The parallel lines *DE*, *FG*, *HJ*, and *KL* are called *factor-cost* or *isocost* curves. If the amount necessary to hire *OD* units of factor *Y* were used to hire factor *X*, *OE* of *X* could be purchased. The slope of *DE* is minus the price of *X* in terms of *Y*, and all points on *DE* represent the same total cost. The greatest total output that can be produced with the total cost represented by *DE* is determined by the point of tangency between *DE* and an isoquant. Or, putting it the other way around, the lowest total cost at which 100 units of output can be produced is by using Ox of X in combination with Oy of Y; this assumes, of course, that there are only two factors used in the production of the product in question.

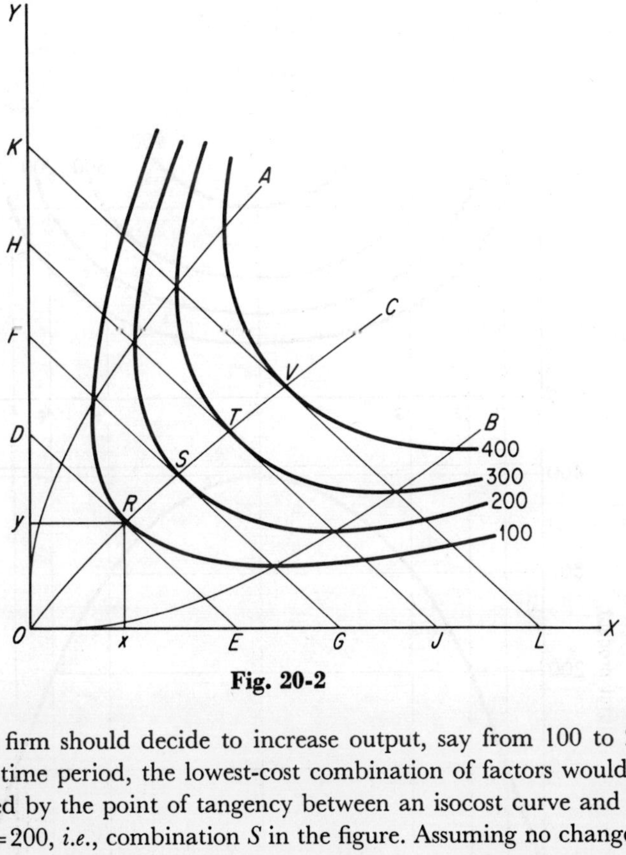

Fig. 20-2

If the firm should decide to increase output, say from 100 to 200 units per time period, the lowest-cost combination of factors would be determined by the point of tangency between an isocost curve and the isoquant = 200, *i.e.*, combination S in the figure. Assuming no change in relative input prices, the lowest-cost combinations of factors as output increases are traced by the isocline OC. This curve is known as the *expansion path*.

20.3 Isoquants and the Total Product Curve

In the upper part of Figure 20-3, we have a set of isoquants representing outputs from 100 to 400 per time period. We assume that Y is a fixed factor, and the firm has OF of Y. The fixed factor may be used in combination with various quantities of X, the variable factor.

If the firm used Ox_1 of X with OF of Y, the total product will be 100 units. In the lower part of the figure we may, accordingly, mark the point 100 total product when $x = Ox_1$. Similarly, when Ox_2 of X is used

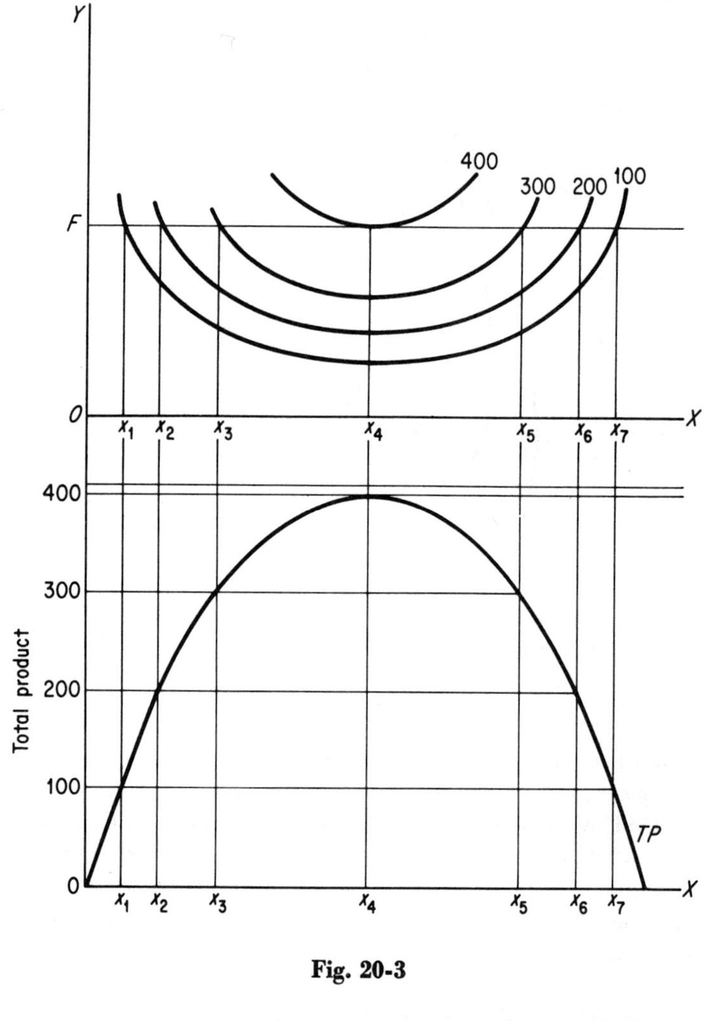

Fig. 20-3

with OF of the fixed factor, the total product is 200. This point may be located in the lower part of the chart. In the same manner we may determine the total product for the several values of x from $x = Ox_1$ to $x = Ox_7$.

20.4 Exercises

1. The table below represents the total output resulting from the various combinations of land and labor. If, for example, 1 unit of labor is used in combination with 3 units of land, the total product is 245.
   (a) Draw the isoquants $q=282$, $q=346$, and $q=490$.
   (b) Assume that land costs \$3 per unit and labor costs \$2 per unit. Draw the isocost curves $TC=\$10$, $TC=\$12$, and $TC=\$17$.
   (c) Draw the expansion path.

Units of Land						
6	346	490	600	693	775	846
5	316	448	548	632	705	775
4	282	400	490	564	632	693
3	245	346	423	490	548	600
2	200	282	346	400	448	490
1	141	200	245	202	316	346
	1	2	3	4	5	6
	Units of Labor					

2. Draw a set of isoquants. Derive from them the total product curve, and from that derive the average and marginal product curves.

20.5 Bibliographical Note

For further discussion of isoquants the student may consult Kenneth E. Boulding, *Economic Analysis*, 3rd ed., Chapter 34; Ralph K. Davidson, Vernon L. Smith, and Jay W. Wiley, *Economics: an Analytical Approach*, Chapters 19–20; John F. Due, *Intermediate Economic Analysis*, 3rd ed., pp. 134–139; Richard H. Leftwich, *The Price System and Resource Allocation*, rev. ed., pp. 126–135; Joseph P. McKenna, *Intermediate Economic Theory*, Chapter 3; Tibor Scitovsky, *Welfare and Competition*, Chapters 6–7; Alfred W. Stonier and Douglas C. Hague, *A Textbook of Economic Theory*, Chapter 10; Sidney Weintraub, *Price Theory*, Chapter 3.

21

The Nature of Linear Programming

21.1 Homogeneous Production Function

We begin with an analysis of a problem in factor allocation using the isoquant technique. Consider the special case of the production function that is homogeneous in the first degree. Such a function is illustrated in Figure 21-1. It will be noted that a given increase in each of the quantities of the two factors employed, X_1 and X_2, will result in a proportionate increase in output. If 2 units of X_1 are used in combination with 4 units of X_2, for example, total output is 100. If the quantity of each factor employed is doubled, the total output will be doubled; with 4 units of X_1 and 8 units of X_2, the total output will be 200. With 6 units of X_1 and 12 units of X_2, the total output is 300, and so on.

If the isocost lines are $R_1R'_1$, $R_2R'_2$, $R_3R'_3$, and so on, the lowest cost at which 100 units of output can be produced is with 2 units of X_1 and 4 units of X_2, *i.e.*, point L in the figure. The lowest-cost combination of factors for an output of 200 is represented by point M in the figure. The expansion path with these factor costs given is represented by OA, which is linear. If, on the other hand, the isocost curves are $P_1P'_1$, $P_2P'_2$, $P_3P'_3$, and so on, the expansion path will be OB, which is, of course, also linear.

The isoquants indicate that the factors X_1 and X_2 may be used in any combination. If, for example, we use 4 of X_1 with 2 of X_2, the total product will be 100. If we use 4 of X_1 with 8 of X_2, the total product will be 200. Suppose now that the amount of factor X_2 is fixed at 8; the firm can use any quantity of factor X_1 in combination with 8 of X_2. If 1 of X_1 is used with 8 of X_2, the total product will be 100. If 4 of X_1

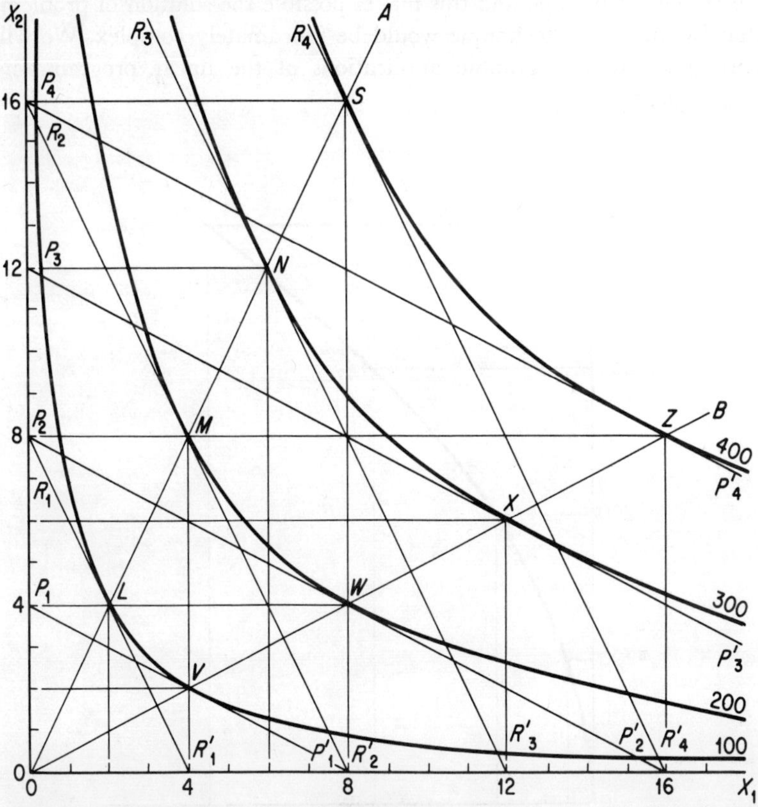

Fig. 21-1

are used with 8 of X_2, the total product will be 200. With 9 of X_1 and 8 of X_2, the total product will be 300. And 16 of X_1 plus 8 of X_2 will result in a total product of 400. These points have been plotted in Figure 21-2 indicating the shape of the total product curve.

21.2 Linear Programming and Optimum Combination of Factors

In recent years the use of a technique variously known as linear programming, activity analysis, and mathematical programming has proved to be a powerful tool in the analysis of the problems of resource allocation at a variety of levels. Linear programming has a number of advantages over alternate techniques; in many applications it is essentially simpler and requires less information than the marginal analysis. In algebraic formulations it makes possible an effective utilization of electronic calculators, and this makes possible the solution of problems that by any other technique would be inordinately complex. We will here illustrate two graphic applications of the linear programming technique.

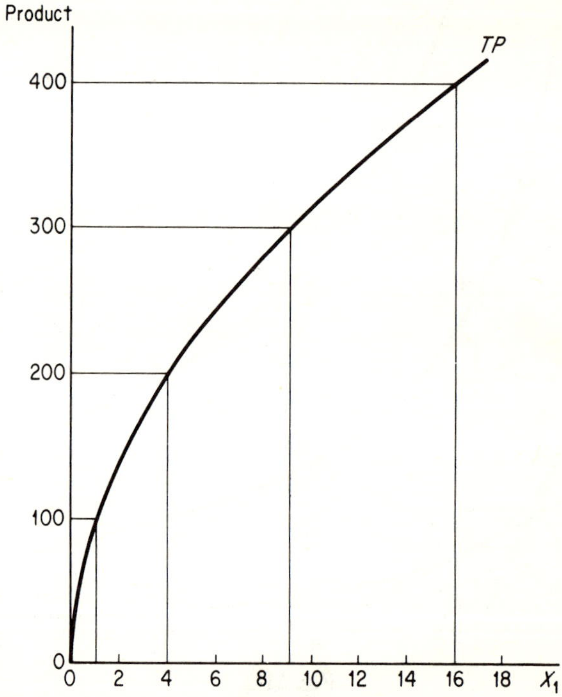

Fig. 21-2

We turn our attention first to the problem of the optimum production technique as it might be solved by linear programming in contrast to the method of isoquants, which we have illustrated in the preceding section. We make a number of critical simplifying assumptions.

1. First, we assume that there are several processes by which factors may be combined to attain any given output. In Figure 21-3, *OA* represents one process and *OB* represents another. If we use process *A*, we can get a total product of 100 by employing 2 units of X_1 and 4 units of X_2, a total product of 200 by using 4 units of X_1 and 8 units of X_2, and so on. If we use process B, we can get a total product of 100 by using 4 units of X_1 and 2 units of X_2, a total product of 200 by employing 8 units of X_1 and 4 units of X_2, and so on. The process rays are essentially identical to the expansion paths of Figure 21-1.

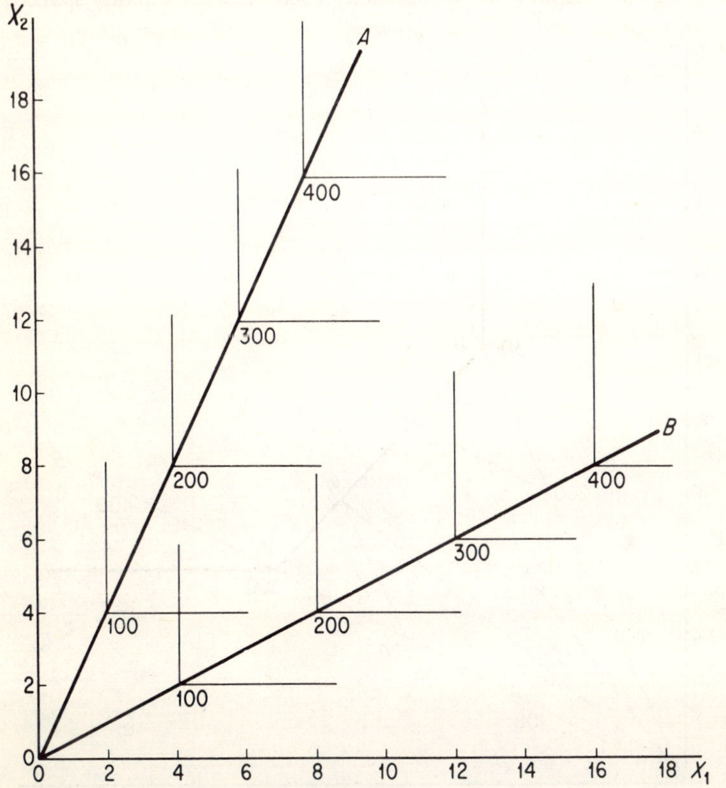

Fig. 21-3

2. Then we assume that, in each process, factors are used in fixed proportions, *e.g.*, two men to a machine or one shovel to each man. At each output point indicated on the process rays in Figure 21-3 a 90° "isoquant" has been sketched in. With process A we can, for example, produce 200 units of output with 4 units of X_1 and 8 units of X_2. If we used more than 4 units of X_1 with 8 units of X_2, the output would remain at 200, and all of the units in excess of 4 would be redundant. The geometric implication of the fixed-proportion assumption is that the process rays have a constant slope; *i.e.*, they are linear.

3. Next, we assume that the processes are homogeneous in the first degree, or, as it may be expressed, we assume constant returns to scale. If, for example, we double the amount of each factor employed, the output will be doubled. This is reflected by the fact that the points representing the output scale along each process ray are equally spaced.

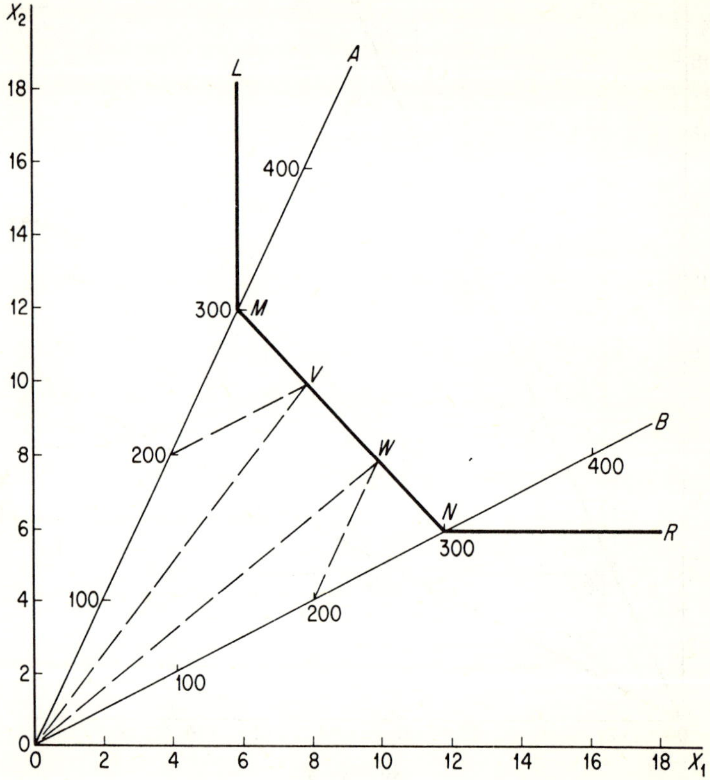

Fig. 21-4

4. Finally, we assume that the use of two processes simultaneously does not affect the efficiency of either. In Figure 21-4, we have reproduced the process rays of Figure 21-3 and have drawn in the isoproduct contour *LMNR*, which indicates the combinations of X_1 and X_2 that will result in a total product of 300. We know that by using process *A*, we can produce 300 units of output by employing 6 units of X_1 with 12 units of X_2—point *M* on the isoproduct contour; or by using process *B*, we can get a total product of 300 by employing 12 of X_1 and 6 of X_2—

point *N*. But it would also be possible to get a total product of 300 by using processes *A* and *B* in combination. We might, for example, produce 200 units of output by process *A* and 100 units by process *B*. This would in effect provide us with a new process, *OV*, which would require

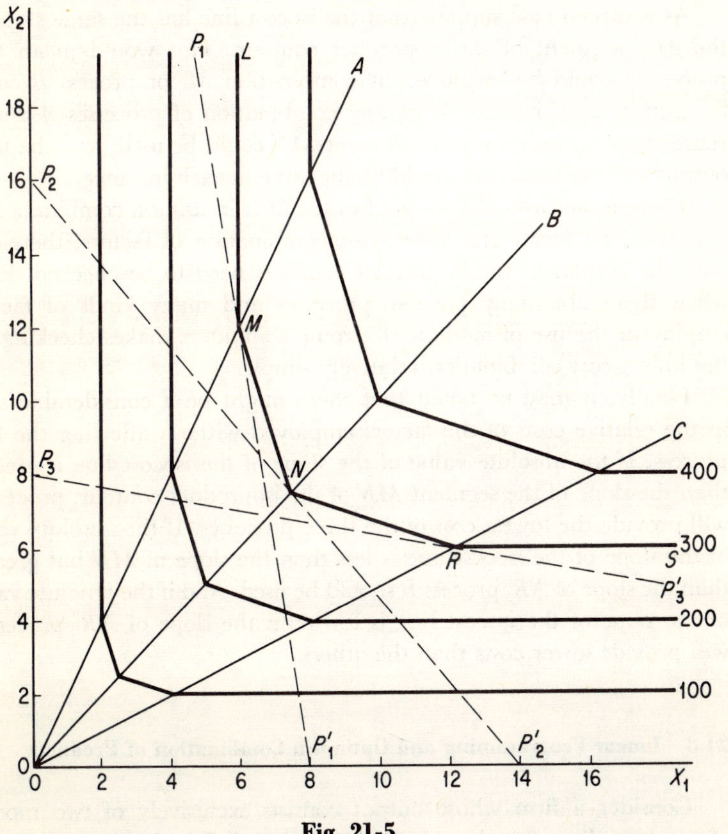

Fig. 21-5

a total of 8 units of factor X_1 and 10 units of X_2. Or, we might get a total of 300 units of output by producing 200 by process *B* and 100 by process *A*; this would be process *OW*, requiring 10 units of X_1 and 8 units of X_2.

The choice of the lowest-cost process will, of course, depend on the relative costs of the factors X_1 and X_2; *i.e.*, on the slopes of the isocost contours. If in Figure 21-5 the costs of the factors are reflected in the slope of the isocost line $P_1P'_1$, process *A* will be the best process; 300 units of output, for example, can be produced at the lowest cost with combination M; *i.e.*, 6 units of X_1 and 12 units of X_2. If, however, $P_2P'_2$ is the appropriate isocost contour, combination N by process B will result in the lowest possible unit costs. And if the isocost line is $P_3P'_3$, process C at combination R will result in an output of 300 units at the lowest cost.

As a special case suppose that the isocost line has the same slope as the MN segment of the isoproduct contour. This would mean that process A could be employed at combination M, or process B could be used at combination N, or any combination of processes A and B represented by points on the segment MN could be used, and the total output and the total cost would be the same in each instance.

There is, accordingly, no gain to the firm in using a combination of processes. To locate the lowest-cost combination of factors, therefore, only the corners of the isoproduct contours need to be checked. Even when there are many possible processes and many kinds of factors employed, the use of modern electronic computers makes checking for the lowest-cost combination relatively simple.

Finally, it may be noted that there might be a considerable shift in the relative costs of the factors employed without affecting the best process. If the absolute value of the slope of the isocost line is greater than the slope of the segment MN of the isoproduct contour, process A will provide the lowest cost of the three processes. If the absolute value of the slope of the isocost line is less than the slope of MN but greater than the slope of NR, process B should be used. And if the absolute value of the slope of the isocost line is less than the slope of NR, process C will provide lower costs than the others.

21.3 Linear Programming and Optimum Combination of Products

Consider a firm whose output consists exclusively of two models of a commodity: S is the standard model and D is the de luxe model. Two machines, M and N, are employed in the production of each model. One unit of S requires M for 2 hours and N for $\frac{4}{3}$ hours. One unit of D requires M for $\frac{3}{2}$ hours and N for 2 hours. The local union requires that no machine is to work more than 12 hours per day. We assume that the firm can sell as many of each product at the going price as it can produce, and the profit on each unit of S is \$3 and on each unit of D is \$4 regardless of the number produced. The problem is to determine how many units of each product should be produced in order to maximize profits.

Let $x =$ the number of units of S produced
and $y =$ the number of D produced.

The total profit will then be

$$P = 3x+4y. \qquad (1)$$

Equation (1) is known as the *criterion* function; we must find x and y, which will maximize P subject to the following constraints:

$$x \geqq 0 \qquad (2)$$
$$y \geqq 0 \qquad (3)$$
$$2x+\tfrac{3}{2}y \leqq 12 \qquad (4)$$
$$\tfrac{4}{3}x+2y \leqq 12 \qquad (5)$$

Constraints (2) and (3) reflect the fact that we cannot produce negative quantities of either product. Constraints (4) and (5) indicate that the machines cannot be used for more than 12 hours per day.

In Figure 21-6, the function $2x+\frac{3}{2}y=12$ has been plotted and labeled MM'. Any point *on* or *below* MM' represents a *feasible* point; that is, total output may be equal to or less than the capacity of machine M, but it cannot exceed the machine's capacity. The function $\frac{4}{3}x+2y=12$ is plotted as NN' and indicates feasible points in terms of the use of machine N. The shaded area, $ONLM'$, represents the area of feasible points subject to the capacity limitations of both machines; any point on or below NLM' represents a production-possibility combination within the constraints of the problem.

The remaining problem is to find the *optimal* feasible point in the production-possibility region. It will be noted that the profit function,

$$P = 3x+4y,$$

has a slope of $-\frac{3}{4}$. A number of isoprofit contours with this slope have been drawn in the figure as broken lines. The higher the isoprofit contour, the greater the profit. Our task is to find the point within the production-possibility region which lies on the highest isoprofit contour. This is seen to be at the corner point L. The most profitable combination of products will be, accordingly, 3 units of model S and 4 units of model D. Total profit will be

$$P = 3\cdot \$3+4\cdot \$4 = \$25.$$

No other combination of products would produce this much profit.

At combination L both machines are employed full time. To produce 3 units of S requires $3\times 2=6$ hours of machine M, and 4 units of D require $4\times\frac{3}{2}=6$ hours of machine M, a total of 12 hours. To produce 3 units of S requires $3\times\frac{4}{3}=4$ hours of machine N, and 4 units of D require $4\times 2=8$ hours of machine N, a total of 12 hours. If, however, as the student can verify for himself, the profit on each unit of S had

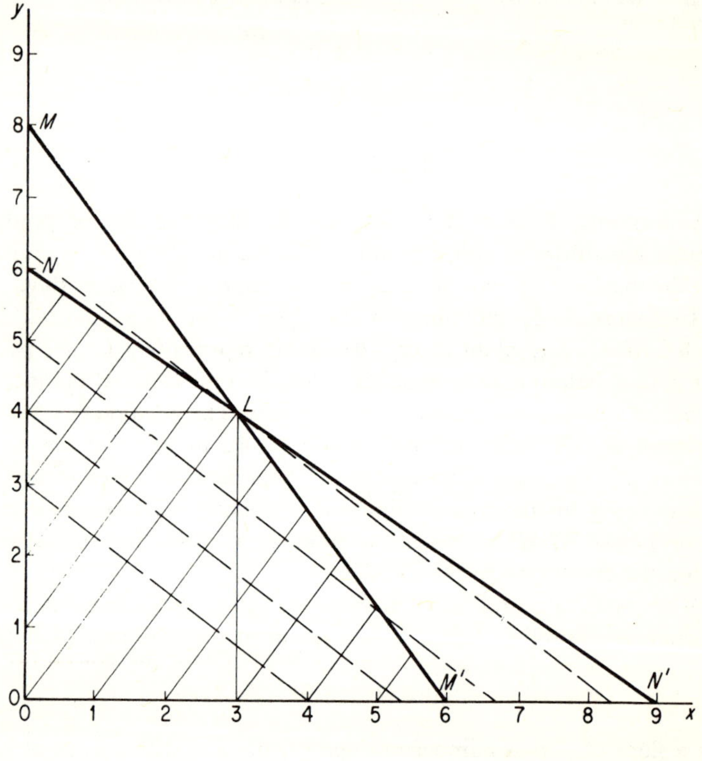

Fig. 21-6

been \$3 and on each unit of D had been \$5, the best combination of products would have been at point N; that is, the firm then should produce 6 units of D and none of S. To produce 6 units of D would require $2\times 6=12$ hours of machine N, but only $6\times\frac{3}{2}=9$ hours of machine M. Machine M would, accordingly, be only partially employed. But total profit at this output would be $6\times\$5=\$30$, and no other combination of outputs would result in a profit so great.

21.4 Exercises

1. (a) Plot the following three isoquants:

$$xy = 12 \tag{1}$$
$$xy = 48 \tag{2}$$
$$xy = 108 \tag{3}$$

(b) Let isoquant (1) represent a total output of 50, isoquant (2) an output of 100, and isoquant (3) an output of 150. Show that this is a homogeneous production function of the first degree.

(c) Suppose that by process A the production of 50 units of output requires the use of 3 units of factor X and 4 units of factor Y. How much X and Y will be required to produce 100 units of output by process A? to produce 150 units of output?

(d) Draw the expansion path for process A and label it OA.

(e) What must be the price of factor X in terms of the price of factor Y in order to make process A the lowest-cost process?

(f) Suppose that by process B it requires 4 units of X and 3 units of Y to produce 50 units of output. Draw the expansion path of process B and label it OB.

(g) What must be the price of factor X in terms of the price of factor Y to make process B the lowest-cost process?

2. Assume that in the preceding problem factor Y is fixed at 6. Draw the appropriate total product curve.

3. Consider a production function that is homogeneous in the first degree. Two factors, X_1 and X_2 are employed, and either of two processes may be employed.

(a) If process A is used, it requires 20 units of X_1 and 30 units of X_2 to produce an output of 200. Draw the process ray OA and mark off on it the output scale from 100 to 500.

(b) If process B is employed, it requires 40 of X_1 and 20 of X_2 to produce an output of 200. Draw the process ray OB and mark off the output scale from 100 to 500.

(c) Draw the linear isoproduct contour representing an output of 400.

(d) If each unit of X_2 costs three times as much as a unit of X_1, which process should be used to minimize costs, and how much of each factor should be employed to produce 400 units?

(e) If the cost of a unit of X_2 is double the cost of a unit of X_1, which process should be used?

(f) If both factors come at the same price, which is the lowest-cost process?

4. Suppose that a person wants to subsist at minimum cost on only two foods, bread and cheese. The minimum nutrient requirements are 3,000 calories and 100 grams of protein per day. The nutrient contents of the foods are 1,000 calories and 25 grams of protein per pound of bread and 2,000 calories and 100 grams of protein per pound of cheese. The market prices are 24¢ per pound of bread and 84¢ per pound of cheese. The essential data are summarized in the following table:

	One unit of		
	B	*C*	Requirements
Price (¢ per unit)	24	84	
Calories (1,000)	1	2	3
Protein (25 gr.)	1	4	4

Solve this linear program graphically, indicating (a) the amount of bread, (b) the amount of cheese, and (c) the total minimum cost of the diet that meets the constraints imposed.

21.5 Bibliographical Note

The student is urged to consult what has come to be virtually the standard introduction to the study of linear programming: Robert Dorfman, "Mathematical or 'Linear' Programming: a Non-Mathematical Exposition," *American Economic Review*, December 1953, pp. 797–825. Several excellent textbooks in linear programming have appeared in recent years; among them are the following: Kenneth E. Boulding and W. Allen Spivey, *Linear Programming and the Theory of the Firm*; Robert Dorfman, Paul A. Samuelson, and Robert M. Solow, *Linear Programming and Economic Analysis*; and Helen Makower, *Activity Analysis and the Theory of Economic Equilibrium.* Short introductions to the subject may be more useful to the student at this point and include: R. G. D. Allen, *Mathematical Economics*, pp. 535–541; Ralph K. Davidson, Vernon L. Smith, and Jay W. Wiley, *Economics: an Analytical Approach*, Chapter 21; Joseph P. McKenna, *Intermediate Economic Theory,* Chapter 4; M. Richardson, *Fundamentals of Mathematics*, rev. ed., pp. 259–265.

Appendixes

A

Some Fundamentals Reviewed

In presenting the mathematical tools needed by nonmathematical economists it would seem appropriate to begin at the beginning. However, since all of us began to make use of mathematical concepts almost as soon as we began to talk, it is hardly practicable to begin at the *very* beginning. Before reaching the age of two a child is likely to distinguish between "one cow, two cows, and many cows." It appears to be a safe assumption that college students have acquired the ability to add, subtract, divide, and multiply (although cases of college students who think that $2 \times \frac{1}{2} = \frac{2}{4} = \frac{1}{2}$ are not uncommon). In short, we cannot literally start at the beginning, and some mathematical knowledge on the part of the student has to be assumed.

The hazard is, of course, that we may assume such a degree of mathematical sophistication that those who need this sort of book will be unable to use it, and those who can use it will not need it. And since in a single class a variety of mathematical backgrounds will be represented, there is no completely satisfactory starting point. To fill in gaps and refresh memories, a number of points that the student once knew but may have long since forgotten are summarized in this appendix. We will not be primarily concerned with proofs or demonstrations, and there will be no exercises to test the student's understanding. For some, what is given here may prove to be too sketchy, and a perusal of appropriate high school or college mathematics textbooks may be desirable. And there may be points not included that some students may need to learn or relearn as the course progresses. What is presented here may be taken as the minimum essentials, and this plus what the student already knows will generally provide a sufficient background for the course.

1. The Fundamental Operations of Addition and Subtraction

(1) The sum of two numbers is a uniquely determined number. Given a and b, there is one and only one number x to give $a+b=x$.

(2) Addition is commutative. That is,

$$a+b = b+a.$$

(3) Addition is associative. That is,

$$(a+b)+c = a+(b+c).$$

(4) If equal numbers are added to equal numbers, the sums are equal numbers.

If

$$a = b$$

and

$$c = d,$$

then

$$a+c = b+d.$$

(5) The product of any two numbers is a uniquely determined number. Given a and b, there is only one number y to give $ab=y$. In this case a and b are said to be *factors* of y.

(6) Multiplication is commutative. That is,

$$ab = ba$$

(7) Multiplication is associative. That is,

$$(ab)c = a(bc).$$

(8) Multiplication is distributive with respect to addition. That is,

$$a(b+c) = ab+ac.$$

(9) If equal numbers are multiplied by equal numbers, the products are equal. That is, if

$$a = b$$

and

$$c = d,$$

then

$$ac = bd.$$

(10) Given a and b, there is one and only one number x to give $x+b=a$.

If $a=b$ and $x+b=a$, $x=0$.
If $x+b=0$ and b is a positive number, x is a *negative* number.

That is,

$$(-b)+b = 0,$$

and this defines $(-b)$.

(11) Given a and b ($b \neq 0$, *i.e.*, b is not equal to 0), there is one and only one number x to give $bx=a$.

Division is the process of finding the number x in $bx=a$.
In dividing a by b, the number a is called the *dividend* and b the *divisor*.

In the fraction $\frac{a}{b}$, a is called the *numerator* and b the *denominator*.

The number x in

$$bx = a \qquad (b \neq 0)$$

exists when $b=a$, so that $ax=a$. In this case the number x is called *unity*; that is,

$$\frac{a}{a} = 1.$$

Further, there exists a number x that satisfies

$$bx = 1 \qquad (b \neq 0).$$

This value of x is called the *reciprocal* of b and is written $\frac{1}{b}$.

2. Derived Properties of the Numbers of Algebra

(1) Adding a negative number $(-a)$ is equivalent to subtracting a positive number a. That is,

$$b+(-a) = b-a.$$

(2) Subtracting a negative number $(-a)$ is equivalent to adding a positive number a. That is,

$$b-(-a) = b+a.$$

(3) The product of two numbers is 0 when and only when at least one of the numbers is 0.

As a corollary it may be said that the quotient $\frac{0}{a}$ is equal to 0 when a is any number other than 0.

(4) The product of a positive number a and a negative number $(-b)$ is $-ab$.

(5) The product of $(-a)$ and $(-b)$ is ab.

(6) The quotient of two numbers is positive if the signs of the dividend and divisor are alike; negative if they are unlike.

(7) A single parenthesis may be removed when it is preceded by a positive sign without changing the signs of the terms within it.

(8) A single parenthesis may be removed when it is preceded by a negative sign if the signs of the terms within it are changed. That is,

$$-(a+b-c+d) = -a-b+c-d.$$

(9) The value of a fraction is not changed by multiplying or dividing both the numerator and denominator by the same number. That is,

$$\frac{a}{b} = \frac{ax}{bx} \qquad (x \neq 0).$$

(10) Changing the sign of either the numerator or the denominator of a fraction is equivalent to changing the sign of the fraction. That is,

$$\frac{-a}{b} = \frac{a}{-b} = -\frac{a}{b}.$$

(11) Adding two fractions having a common denominator gives a fraction whose numerator is the sum of the numerators and whose denominator is the common denominator. That is,

$$\frac{a}{b} + \frac{c}{b} = \frac{a+c}{b}.$$

(12) The sum and the difference of two fractions are expressed by

$$\frac{a}{b} + \frac{c}{d} = \frac{ad+bc}{bd}, \quad \text{and} \quad \frac{a}{b} - \frac{c}{d} = \frac{ad-bc}{bd},$$

respectively.

(13) The product of two fractions is a fraction whose numerator is the product of the numerators and whose denominator is the product of the denominators. That is,

$$\frac{a}{b} \cdot \frac{c}{d} = \frac{ac}{bd}.$$

(14) To divide one fraction by another, invert the latter and then multiply one by the other. That is,

$$\frac{a}{b} \div \frac{c}{d} \quad \text{or} \quad \frac{\frac{a}{b}}{\frac{c}{d}} = \frac{a}{b} \cdot \frac{d}{c} = \frac{ad}{bc}.$$

3. Rules of Exponents

(1) $a^m a^n = a^{m+n}$.
For example, $5^3 \cdot 5^4 = 5^7$.

(2) $(a^m)^n = a^{mn}$.
For example, $(3^4)^5 = 3^{20}$.

(3) $(abc \cdots)^m = a^m b^m c^m \cdots$.
For example, $(3 \cdot 4 \cdot 5)^2 = 3^2 \cdot 4^2 \cdot 5^2$.

(4) $\left(\frac{a}{b}\right)^m = \frac{a^m}{b^m}$.
For example,

$$\left(\frac{3}{7}\right)^3 = \frac{3^3}{7^3}.$$

(5) $\frac{a^m}{a^n} = a^{m-n}\ (m > n)$.
For example,

$$\frac{5^6}{5^4} = 5^2.$$

(6) $\frac{a^m}{a^n} = \frac{1}{a^{n-m}}\ (m < n)$.

For example,

$$\frac{5^4}{5^6} = \frac{1}{5^2}.$$

(7) $a^{-m} = \frac{1}{a^m}\ (a \neq 0)$.
For example,

$$6^{-3} = \frac{1}{6^3}.$$

(8) $a^0 = 1$.
For example, $100^0 = 1$.

(9) $a^{1/q} = \sqrt[q]{a}$ and is read the qth root of a.
For example,

$$8^{1/3} = \sqrt[3]{8}.$$

(10) $a^{p/q} = (\sqrt[q]{a})^p$ and is read the pth power of the qth root of a. For example,

$$100^{3/4} = (\sqrt[4]{100})^3.$$

4. Addition and Multiplication of Polynomials

Algebraic expressions may be added or multiplied in much the same way that numbers in arithmetic are manipulated. Suppose, for example, that we want to determine the sum of the two expressions

$$(4x^2+6x+10)+(2x^2-2x-15).$$

We may write "like powers" under each other and proceed as follows:

$$\begin{array}{r} 4x^2+6x+10 \\ \underline{2x^2-2x-15} \\ 6x^2+4x-\ \ 5. \end{array}$$

Similarly, the polynomials $(4x+4)$ and $(3x-2)$ may be multiplied as follows:

$$\begin{array}{r} 4x+4 \\ \underline{3x-2} \\ -8x-8 \\ \underline{12x^2+12x} \\ 12x^2+\ 4x-8. \end{array}$$

5. Factoring

The process of starting with a product and determining what factors were multiplied to get it is called factoring. This process is usually more difficult than multiplication, and only the simplest cases are reviewed here.

(1) Taking out a common factor. For example,

$$6x^2+12x^2y+18xy = 6x(x+2xy+3y).$$

(2) Difference of two squares. Since we know that by multiplication $(a+b)(a-b)=a^2-b^2$, we may factor

$$x^2-y^2 = (x+y)(x-y).$$

(3) Quadratic trinomials. Expressions of the form $x^2-7x+12$ must be factored by trial and error. If this expression can be factored, it will be the product of two factors of the form $(x-?)\ (x-?)$. Two numbers must be found whose algebraic sum is -7 and whose product is $+12$. By experimentation it is found that

$$x^2-7x+12 = (x-3)\ (x-4).$$

6. Equations

(1) If an equation becomes a true statement for every value of x for which the expression involved has a value, it is called an identical equation or an *identity.*

For example, the following are identities:

$$3x(x+y) = 3x^2+3xy \tag{1}$$

$$(x+y)^2 = x^2+2xy+y^2 \tag{2}$$

$$\frac{1}{2x+3}+\frac{1}{x-2} = \frac{3x+1}{(2x+3)(x-2)} \tag{3}$$

(2) If an equation becomes a false statement for some values of x, it is called a conditional equation or simply an equation. A value of x that does satisfy the equation is called a *root* of the equation. For example,

$$3x-12 = 0$$

is a true statement only when x has a value of 4.

Every polynomial equation can be written with a polynomial on the left side of the equal sign and a zero on the right side. For example, if

$$2x^2+5x+6 = x^2+3x+2,$$

then

$$x^2+2x+4 = 0.$$

The highest power of the unknown in an equation determines the *degree* of the equation. If the highest power of x is 1, the equation is linear. For example,

$$6x+36 = 0$$

is a linear equation.

If the highest power of x is 2, the equation is quadratic. For example,

$$4x^2+6x-12 = 0$$

is a quadratic equation.

If the highest power of x is 3, the equation is cubic. For example,

$$x^3+2x^2+5x+7 = 0$$

is a cubic equation.

A quadratic equation may be represented by the general form

$$ax^2+bx+c = 0, \qquad (a \neq 0)$$

where a, b, and c are constants. When the roots of a quadratic equation cannot readily be determined by the simpler factoring rules, they may be determined by the following formula:

$$x = \frac{-b \pm \sqrt{b^2-4ac}}{2a}.$$

7. Simultaneous Equations

Consider a system of two equations such as

$$x+y = 8 \qquad (1)$$

$$x-y = 2. \qquad (2)$$

We find a common solution to these equations by subtracting equation (2) from equation (1):

$$\begin{aligned} x+y &= 8 \\ \underline{x-y} &\underline{= 2} \\ 2y &= 6 \\ y &= 3 \quad \text{and} \quad x = 5. \end{aligned}$$

The values $x=5$ and $y=3$ satisfy both equations.

An alternative method of solving the equations is to write equation (1) as

$$y = 8-x$$

and to substitute this for y in equation (2):

$$\begin{aligned} x-(8-x) &= 2 \\ x-8+x &= 2 \\ 2x &= 10 \\ x &= 5 \quad \text{and} \quad y = 3. \end{aligned}$$

Consider next a system of equations in which one equation is linear and the other is quadratic, such as

$$x^2+y^2 = 25 \tag{3}$$
$$2x+y = 10. \tag{4}$$

From equation (4) we determine that

$$y = 10-2x,$$

and this may be substituted for y in equation (3):

$$x^2+(10-2x)^2 = 25$$
$$x^2+100-40x+4x^2 = 25$$
$$5x^2-40x+75 = 0$$
$$(5x-25)(x-3) = 0$$
$$x = 5; y = 0$$

or

$$x = 3; y = 4.$$

8. Graphic Solutions of Simultaneous Equations

For many purposes it is convenient to represent data graphically. The most common way of doing this is with the use of rectangular coordinates. In Figure A-1, the lines $X'X$ and $Y'Y$ are called *coordinate*

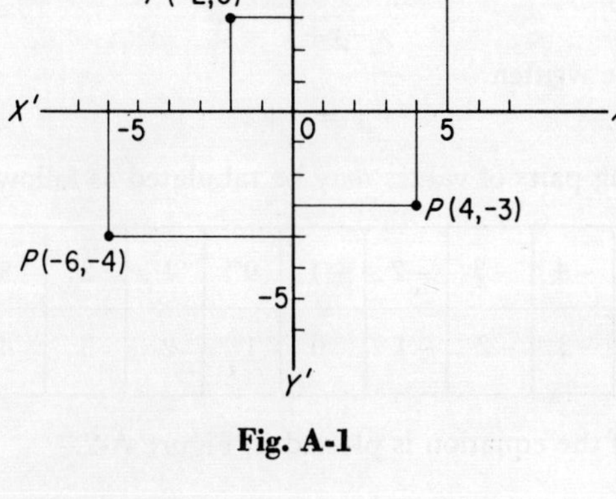

Fig. A-1

axes, and their intersection is called the *origin*. $X'X$ is the x-axis, and $Y'Y$ is the y-axis. The perpendicular distance from a given point to the x-axis is called the *ordinate* or y-value of the point. The perpendicular distance from a given point to the y-axis is called the *abscissa* or x-value of the point. Points above the x-axis have positive y-values, and those below the x-axis have negative y-values. Points to the right of the y-axis have positive x-values, and those to the left of the y-axis have negative x-values.

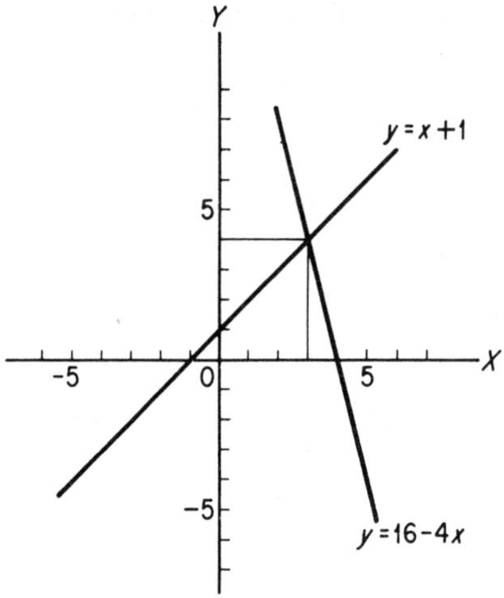

Fig. A-2

The graph or locus or curve of an equation in two variables is the set of all those points and only those points whose coordinates satisfy the equation. Consider the equation

$$x - y + 1 = 0, \tag{1^a}$$

which may be written

$$y = x + 1. \tag{1^b}$$

Corresponding pairs of values may be tabulated as follows:

x	-5	-4	-3	-2	-1	0	1	2	3	4	5
y	-4	-3	-2	-1	0	1	2	3	4	5	6

The graph of the equation is plotted in Figure A-2.

Now consider the equation

$$4x+y-16 = 0, \qquad (2^a)$$

which may be written

$$y = 16-4x. \qquad (2^b)$$

Corresponding pairs of values may be tabulated as follows:

x	-5	-4	-3	-2	-1	0	1	2	3	4	5
y	36	32	28	24	20	16	12	8	4	0	-4

The graph of this equation is also plotted in Figure A-2.

It will be noted that the graphs of the two equations intersect at point $x=3$ and $y=4$. This is written $P(3, 4)$ with the x-value always indicated first. The solution of the two simultaneous equations is, accordingly, $x=3$ and $y=4$.

9. Trigonometric Functions

Consider the right triangle ABC in Figure A-3. Side c is the *hypotenuse*. From the point of view of angle A, side a is the *opposite* side and side b is the *adjacent* side. The trigonometric functions or the trigonometric ratios are defined as follows:

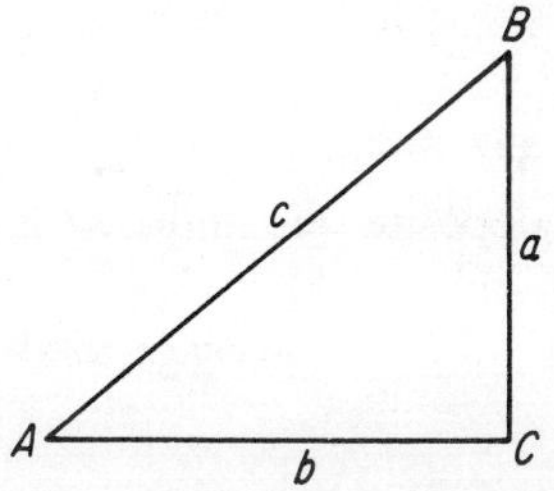

Fig. A-3

$$\sin A = \frac{a}{c} = \frac{\text{opp.}}{\text{hyp.}} \qquad \csc A = \frac{c}{a} = \frac{\text{hyp.}}{\text{opp.}}$$

$$\cos A = \frac{b}{c} = \frac{\text{adj.}}{\text{hyp.}} \qquad \sec A = \frac{c}{b} = \frac{\text{hyp.}}{\text{adj.}}$$

$$\tan A = \frac{a}{b} = \frac{\text{opp.}}{\text{adj.}} \qquad \cot A = \frac{b}{a} = \frac{\text{adj.}}{\text{opp.}}$$

The abbreviations stand for *sine*, *cosine*, *tangent*, *cosecant*, *secant*, and *cotangent* of A.

In economic analysis frequent use is made of the tangent of an angle.

In Figure A-4, C is a total cost curve. At output OM total cost is equal to ML. Average total cost is equal to total cost divided by output, or

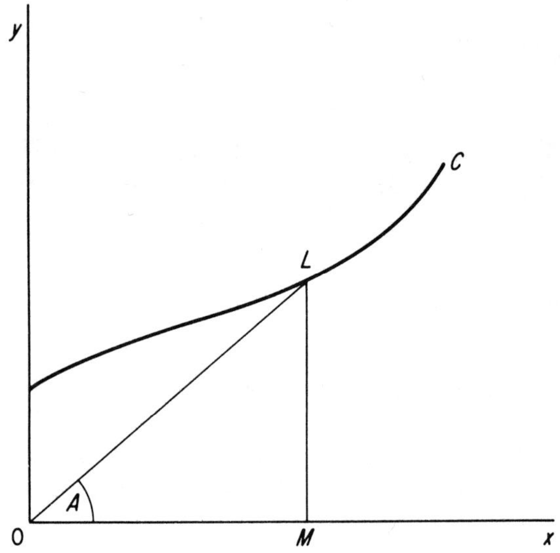

Fig. A-4

ML divided by OM. But from the point of view of angle A, ML is the opposite side and OM is the adjacent side. Accordingly, we may write

$$\text{average total cost} = \frac{ML}{OM} = \tan A.$$

B

Nomographic Solutions of Economic Problems

Economists have made but limited use of geometric solutions of simple equations of the following forms:

$$uv = w \tag{1a}$$

$$\frac{w}{v} = u \tag{2a}$$

$$uvw = q \tag{3a}$$

$$q = \frac{uv}{w} \tag{4a}$$

$$\frac{uv}{wy} \div z = u_1 \tag{5a}$$

$$u + w + v = q \tag{6a}$$

$$\frac{u+v}{w} = z \tag{7a}$$

Science and technology have made extensive use of the geometric method in the solution of problems of these types, particularly in the preparation of alignment charts and other nomograms, and there appear to be a number of applications of these techniques in economic theory. Nomograms make possible the solution of problems in several variables on a two-dimensional chart. The results are frequently interesting and sometimes amusing. In any event, it can scarcely be denied that this is the sort of thing that will interest those who are interested in this sort of thing. The student should try to determine the basic mathematical principle that makes each nomogram work and recall where these principles have been used in the main part of the text.

Consider Figure B-1. The vectors are drawn so that at any u-value

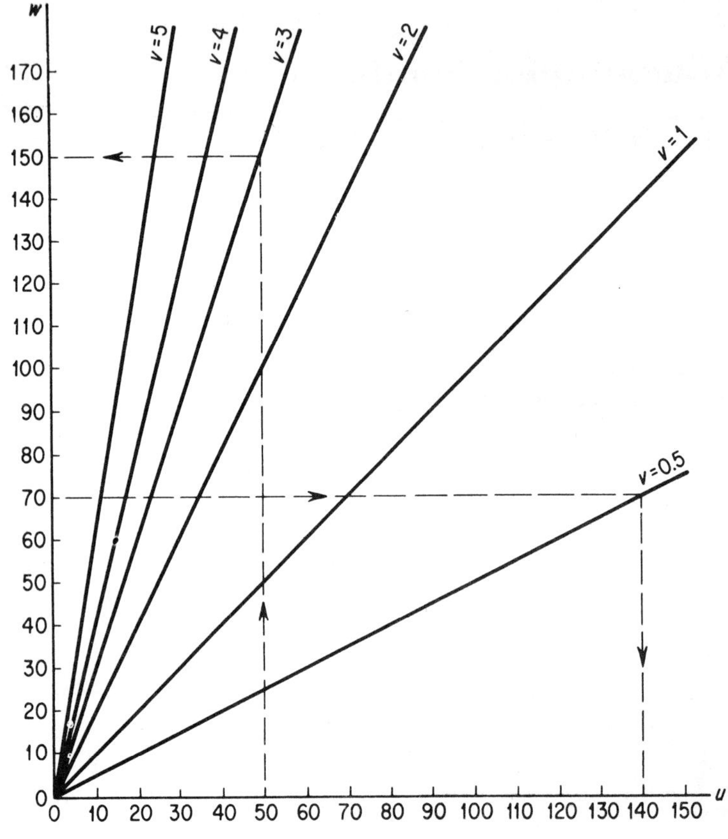

Fig. B-1

the $v=1$ vector has twice the w-value of the $v=0.5$ vector; the $v=2$ vector has twice the w-value of the $v=1$ vector, and so on. To solve an equation of type (1^a) we may assume, for example, that $u=50$ and $v=3$; beginning at $u=50$ and moving in the w direction to vector $v=3$ and then to the left to the w-axis, we read $w=150$.

$$50 \times 3 = 150. \qquad (1^b)$$

To solve an equation of form (2^a) we may assume that $w=70$ and

$v=0.5$. Starting at $w=70$ and moving in the u direction to $v=0.5$ and then moving down to the u-axis, we read $u=140$.

$$70 \div 0.5 = 140. \tag{2^b}$$

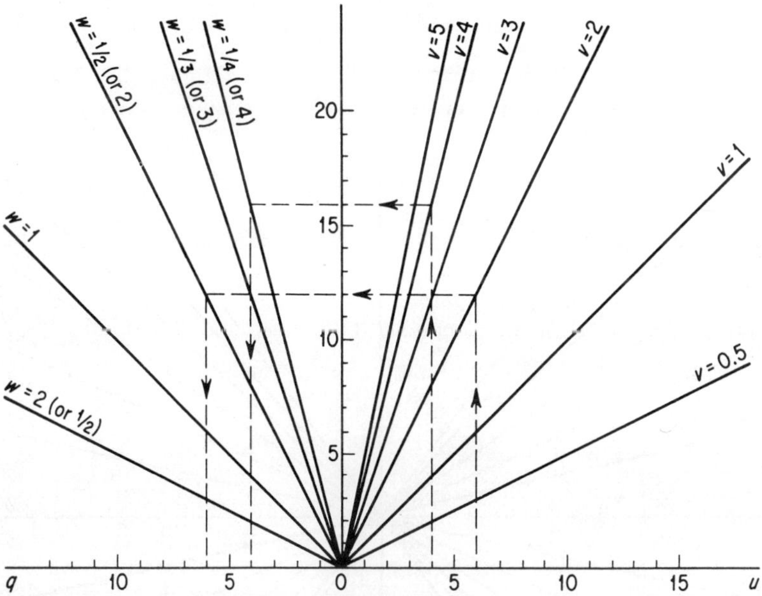

Fig. B-2

Equations with three variables of the form of (3[a]) can be solved with the use of Figure B-2. If we let $u=6$, $v=2$, and $w=0.5$, we read

$$6 \times 2 \times 0.5 = 6. \tag{3^b}$$

Since multiplying by 0.5 is equivalent to dividing by 2, we may renumber the w vectors (in parentheses) so as to divide by w. If we let $u=4$, $v=4$, and $w=4$,

$$4 \times 4 \div 4 = 4. \tag{4^b}$$

This technique may be extended to include a larger number of variables. Figure B-3 may be used to solve equations of form (5[a]). If we let $u=3$, $v=4$, $w=2$, $y=5$, and $z=6$, then

$$3 \times 4 \div 2 \times 5 \div 6 = 5. \tag{5^b}$$

Or, reading in the opposite direction,

$$5 \times 6 \div 5 \times 2 \div 4 = 3. \tag{5c}$$

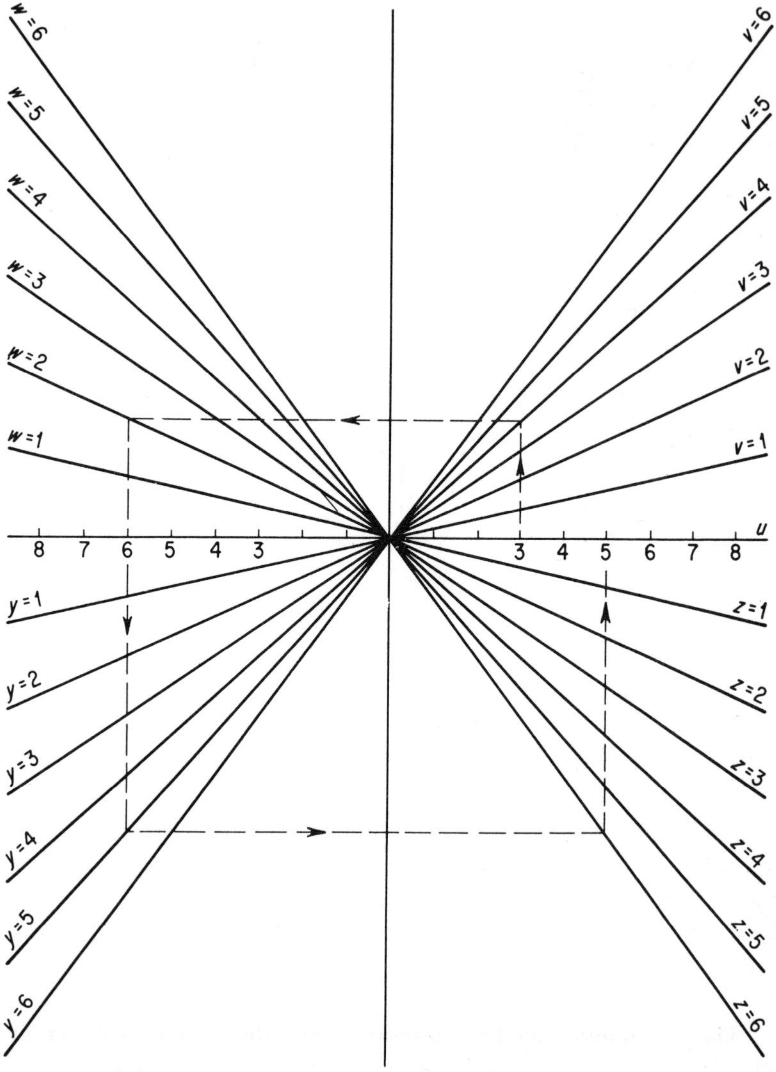

Fig. B-3

Figure B-4 may be used to solve equations of type (6ᵃ), which add or subtract three variables. Suppose that $u=6$, $v=3$, and $w=-4$. Then

$$6+3-4 = 5. \tag{6^b}$$

Or, if $u=8$, $v=-6$, and $w=5$,

$$8-6+5 = 7. \tag{6^c}$$

By utilizing the other two quadrants two additional variables could be handled.

By combining Figures B-2 and B-4 it is possible to solve equations of type (7ᵃ). If $u=6$, $v=4$, and $w=2$, then

$$(6+4) \div 2 = 5. \tag{7^b}$$

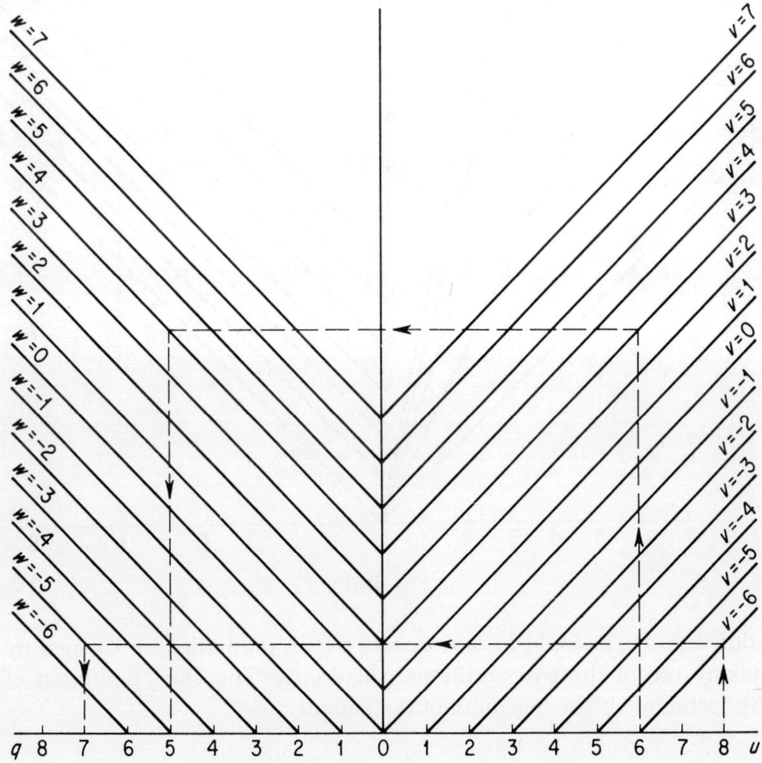

Fig. B-4

If we let u represent fixed cost, v represent variable cost, w represent the quantity produced, and z represent average total cost, Figure B-5 can be used to solve economic problems of the form

$$\frac{FC+VC}{Q} = ATC.$$

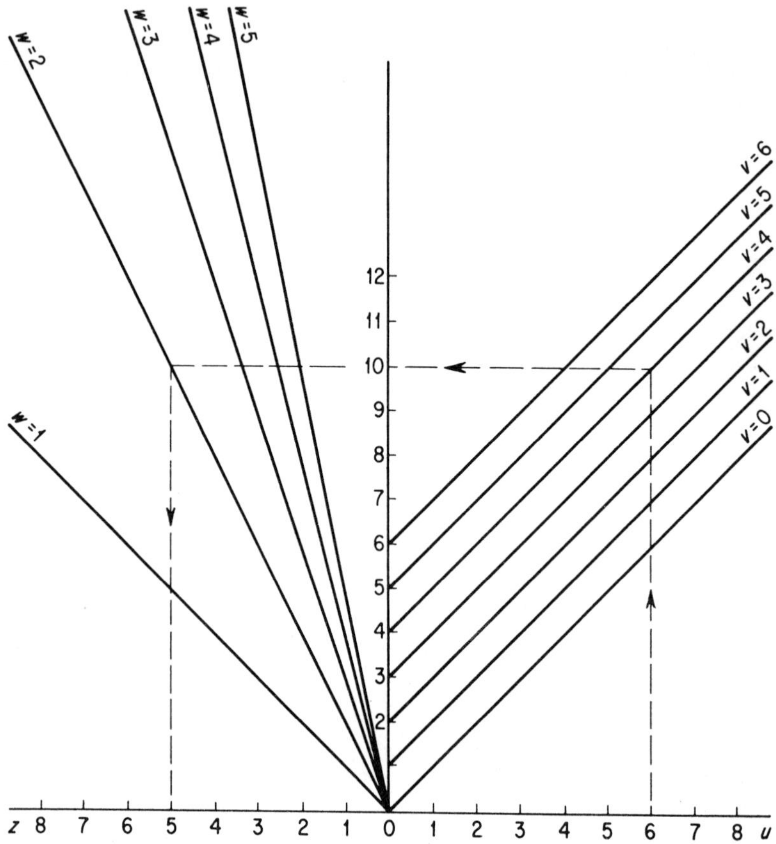

Fig. B-5

Additional terms could be added, subtracted, multiplied, or divided by making use of the two additional quadrants. The chief limitation of this technique is the ingenuity of the designer.

C

Notes on General Equilibrium Theory

1. The Nature of General Equilibrium Theory

We conclude with a brief description of the nature of general equilibrium theory. This analysis will serve two purposes: (a) it will point up the limitations of the familiar partial equilibrium analysis, and (b) it will illustrate a mathematical technique different from any of those discussed in the main body of the text. In this note we will follow the pioneering works of Léon Walras and Gustav Cassel.

One of the most familiar concepts in price theory is the inverse relationship between the price of a commodity and the amount of it that will be demanded. The amount demanded, we say, is a function of price, or

$$D_a = f(p_a).$$

This is represented graphically by a negatively inclined demand curve. But this very useful formulation makes what is known as a *ceteris paribus* assumption; that is, we assume everything else remains the same. Now this is a perfectly proper and, indeed, often necessary kind of assumption to make. When a chemist conducts an experiment, he is careful to see that other things remain the same; he does not want changes in temperature or humidity or air pressure to distort the results of his experiment. But the fact is that if the price of commodity A falls, *everything* else cannot remain the same. If the price of A falls and the prices of all other commodities and the consumer's income remain the same, the consumer's real income has increased. Or, if the price of A falls and the prices of all other commodities and the consumer's real income remain the same, the consumer's money income must fall. Or, if the price of A falls and the prices of all other commodities remain the same, *relative* prices have changed.

The demand function would be more accurately represented if we wrote

$$D_a = f(P_a, P_b, P_c, \cdots, P_n, Y, T);$$

that is, the amount of A demanded depends on the price of A, the prices of all other commodities, the consumer's income, and the consumer's tastes. It is these "other things" that are assumed to remain constant in the partial equilibrium analysis, and to the extent that they do not or cannot remain the same, partial equilibrium provides only approximate answers.

Although it is true that all prices are interrelated, the relationships between some of them are so remote that they can be neglected without perceptibly affecting the validity of the analysis. A change in the demand for cheese in St. Louis, for example, isn't likely to have any measurable effect on the price of locomotives in Philadelphia or the price of theater tickets in New York. No analysis can take fully into account all price interrelationships, and, indeed, it is the nature of abstract theory to draw from reality only those factors that are significant. It is the object of general equilibrium analysis to take into account more of the significant variables.

Although Walras devised a simple general equilibrium system as early as 1874, most contemporary price theory is of the partial equilibrium sort. General equilibrium systems have a particular interest to us here because they set up general equations that we do not attempt to solve. The pricing problem is solved if we have as many equations as we have unknowns. The student will encounter this method of counting equations and unknowns in much of the current literature on price theory.

2. The Case of Fixed Supply

In what follows we shall employ the following symbols:

Let commodities $= 1, 2, 3, \cdots, n$

Factors of production $= 1, 2, 3, \cdots, r$

D_1 = the demand for commodity 1 per time period

S_1 = the supply of commodity 1 per time period

p_1 = the price of commodity 1

q_1 = the price of factor 1

a_{11} = the quantity of factor 1 needed to produce a unit of commodity 1

a_{nr} = the quantity of factor r needed to produce a unit of commodity n

R_1 = the quantity of factor 1 available per time period.

First consider a simple case of a closed economy where supplies of commodities are fixed for the time period under consideration. Call $S_1, S_2, \cdots, S_n$ the supplies of commodities $1, 2, \cdots, n$ per time period. The quantity of money each consumer spends during the period is assumed as given. The demand of each consumer for n commodities will be fixed when the prices of commodities $1, 2, \cdots, n$ are given.

The demand of each consumer is a function of price. Market demand is the summation of the demands of all the consumers in the market. But the demand of consumers for any particular commodity is also a function of the prices of all of the other commodities that they may buy. Market demand for any particular commodity is, accordingly, a function of the prices of n commodities. Call $D_1, D_2 \cdots, D_n$ the market demand for commodities $1, 2, \cdots, n$ per time period. The demand for the n commodities may be set up as follows:

$$\begin{aligned} D_1 &= f_1(p_1, \cdots, p_n) \\ D_2 &= f_2(p_1, \cdots, p_n) \\ &\cdots\cdots\cdots \\ D_n &= f_n(p_1, \cdots, p_n) \end{aligned} \tag{1}$$

where $p_1, \cdots, p_n$ are the prices of the n commodities.

The equations (1) say that the demand for n commodities is a function of the prices of these commodities.

By assumption the supply of each commodity is fixed and known. If the markets are in equilibrium, the demand for each commodity must equal the supply of that commodity; that is, $D_1 = S_1, D_2 = S_2, \cdots, D_n = S_n$. We may then write

$$\begin{aligned} f_1(p_1, \cdots, p_n) &= S_1 \\ f_2(p_1, \cdots, p_n) &= S_2 \\ &\cdots\cdots\cdots \\ f_n(p_1, \cdots, p_n) &= S_n. \end{aligned} \tag{2}$$

We now have n equations and n unknowns (the n prices), and the system is, accordingly, mathematically soluble.

3. The General Case

Let us now drop the assumption that the supply of each commodity is fixed. Suppose that production is carried on under conditions of perfect competition, where price equals cost of production, and cost is determined by the prices of the factors of production employed.

Assume that population, wants, technology, and resources are fixed. Let $1, 2, \cdots, r$ be the factors of production and $R_1, R_2, \cdots, R_r$ the quantities of factors $1, 2, \cdots, r$ available per time period. With these factors n different commodities are produced.

Let a_{11} designate the quantity of factor 1 needed to produce a unit of commodity 1. To produce a unit of commodity n there is needed $a_{n1} \cdots a_{nr}$ factors of production.

Each factor of production must have one price. We shall later take these prices as the unknowns. For the moment let us assume that the prices of the factors of production are given. We designate q_1 as the price of factor 1; q_r is the price of factor r. Since the prices of the commodities will just equal factor costs, we can calculate the prices of the n commodities:

$$\begin{aligned} a_{11}q_1 + a_{12}q_2 + \cdots + a_{1r}q_r &= p_1 \\ a_{21}q_1 + a_{22}q_2 + \cdots + a_{2r}q_r &= p_2 \\ \cdots\cdots\cdots \\ a_{n1}q_1 + a_{n2}q_2 + \cdots + a_{nr}q_r &= p_n. \end{aligned} \tag{3}$$

Now that we know the prices of the commodities, we can calculate the demand for each commodity:

$$\begin{aligned} D_1 &= f_1(p_1, \cdots, p_n) \\ D_2 &= f_2(p_1, \cdots, p_n) \\ \cdots\cdots\cdots \\ D_n &= f_n(p_1, \cdots, p_n) \end{aligned} \tag{4}$$

The equilibrium condition is:

$$D_1 = S_1, D_2 = S_2, \cdots, D_n = S_n, \tag{5}$$

where $S_1, S_2, \cdots, S_n$ are the quantities of $1, 2, \cdots, n$ commodities produced.

Since we know the quantities of n commodities that are produced, we can derive the demand for the factors of production. To produce 1 unit of commodity 1, assuming constant returns to scale, we need $a_{11}S_1, \cdots, a_{1r}S_1$. Thus we need

$$\begin{aligned} &\text{the quantity} \quad a_{11}S_1 + a_{21}S_2 + \cdots + a_{n1}S_n \quad \text{of factor 1} \\ &\text{the quantity} \quad a_{12}S_1 + a_{22}S_2 + \cdots + a_{n2}S_n \quad \text{of factor 2} \\ &\cdots\cdots\cdots \\ &\text{the quantity} \quad a_{1r}S_1 + a_{2r}S_2 + \cdots + a_{nr}S_n \quad \text{of factor } r. \end{aligned} \tag{6}$$

Equations (6) represent the indirect demand of consumers for the factors of production. Under equilibrium conditions in perfectly competitive markets the demand for factors of production must equal the supply. Therefore:

$$
\begin{aligned}
R_1 &= a_{11}S_1 + a_{21}S_2 + \cdots + a_{n1}S_n \\
R_2 &= a_{12}S_1 + a_{22}S_2 + \cdots + a_{n2}S_n \\
&\cdots\cdots\cdots \\
R_r &= a_{1r}S_1 + a_{2r}S_2 + \cdots + a_{nr}S_n.
\end{aligned} \tag{7}
$$

The S's in equations (7), according to equations (5) and (4), are functions of p, which by substitution in equations (3) can be expressed in terms of q. $[S=D=f(p)=F(q)]$. Equations (7) thus contain as unknowns the prices of the factors of production, $q_1, q_2, \cdots, q_n$. The supplies of the factors, $R_1, R_2, \cdots, R_r$, are given. Thus in (7) we have r equations and r unknowns, and we can determine the prices of the factors. With the prices of the factors known, prices of products can be calculated according to (3). The demand for commodities can be obtained from (4). From equations (5) we can determine how much of each commodity must be produced, and this determines the distribution of factors of production among the various industries. From (6) we can determine how much of each factor is needed, and equations (7) assure that the quantities of the factors needed will be available. The system is, accordingly, in general equilibrium.

D

Answers to Odd-Numbered Problems

Chapter 1

Section 3

1. $f(0)=-1; f(1)=2; f(-1)=-10; f(-2)=-31.$

3. $f(a)=3a; f(b)=3b; f(a+b)=3(a+b); f(a)+f(b)=3(a+b).$

Chapter 2

Section 2

1. $x=y-3$. **3.** $x=-4$. **7.** $x=1, y=2$.

Section 4

1. $y=\frac{1}{6}x+\frac{3}{2}$. **3.** $y=4x$.

Section 6

1. (a) $y=\frac{1}{2}x-4$; (b) $y=-\frac{1}{5}x-5\frac{1}{5}$; (c) $y=4$.

3. (a) x-intercept $=-6$; y-intercept $=4$; $m=\frac{2}{3}$;
(b) x-intercept $=-\frac{2}{3}$; y-intercept $=-\frac{1}{2}$; $m=-\frac{3}{4}$.

Chapter 3

Section 2

1. (a) 1; (b) 2; (c) 4; (d) 10; (e) 15; (f) 20; (g) 25; (h) 5.

3. (a) $a/2b$; (b) $2a/3$; (c) a; (d) a/b.

Section 4

1. (a) 5; (b) 6; (c) 14; (d) 10; (e) 35; (f) 460; (g) 4.

3. (a) 1; (b) 0.818; (c) 0.545; (d) 8.7; (e) 6.5; (f) 4.41; (g) 0.0909.

Section 6

1. (a) 3.6; (b) 1.6. **3.** (a) $p=150.4$, $q=209.5$.

Chapter 4

Section 2

1. (a) $a=10$; $b=10$; $m=-1$; (b) $a=50/3$; $b=25$; $m=-\frac{3}{2}$; (c) $a=100$; $b=100/3$; $m=-\frac{1}{3}$; (d) $a=AB$; $b=B$; $m=-1/A$; (e) $a=9$; $b=18$; $m=-2$.

Section 4

1. (a) $a=-10$; $b=10$; $m=1$; (b) $a=-20$; $b=30$; $m=\frac{3}{2}$; (c) $a=-AB$; $b=B$; $m=1/A$; (d) $a=-50$; $b=\frac{1}{2}$; $m=1/100$; (e) $a=-9$; $b=18$; $m=2$.

3. (a) $-Ax+By+C=0$; (b) $a=C/A$; (c) $b=-C/B$; (d) $m=A/B$.

Chapter 6

Section 4

1. $t=0$, $s=0$; $t=1$, $s=16$; $t=2$, $s=64$; $t=3$, $s=144$; $t=4$, $s=256$; $t=5$, $s=400$.

3. $t=0$, $M=0$; $t=1$, $M=32$; $t=2$, $M=64$; $t=3$, $M=96$; $t=4$, $M=128$; $t=5$, $M=160$.

5. 96. Same.

Section 6

1. $y'=5$. **3.** $y'=3x^2+14x-5$. **5.** $y'=\dfrac{-3}{(3x+2)^2}$.

Chapter 8

Section 2

1. (a) $y'=-2x$; $y''=-2$; $y'''=0$; (b) $y'=6x-2$; $y''=6$; $y'''=0$; (c) $y'=nx^{n-1}$; $y''=(n^2-n)x^{n-2}$; $y'''=(n^3-3n^2+2n)x^{n-3}$; (d) $y'=3ax^2+2bx+c$; $y''=6ax+2b$; $y'''=6a$; (e) $p'=-5$; $p''=0$; $p'''=0$.

Section 4

1. (a) $y'=3x^2-8x+8$; $y''=6x-8$; $y'''=6$; $y^{iv}=0$; (b) $y'=3x^2-8x+8$; (d) 11; (f) 11; (g) 10; (i) 10.

Section 6

1. (a) $y'=0.4x-1.6$; $y''=0.4$; (c) -0.8; (e) 0; (g) 0.8.

3. (a) $\pi=-x^3+2x^2+4x-4$; (b) 2.

5. (a) $f'(x)<0$; $f''(x)<0$; (b) $f'(x)=0$; $f''(x)<0$; (c) $f'(x)>0$; $f''(x)<0$; (d) $f'(x)<0$; $f''(x)>0$; (e) $f'(x)=0$; $f''(x)>0$; (f) $f'(x)>0$; $f''(x)>0$.

Chapter 9

Section 2

1. $\partial z/\partial x=4x$; $\partial z/\partial y=-3$. **3.** $\partial z/\partial x=a$; $\partial z/\partial y=b$.

5. $\partial z/\partial x = \dfrac{d}{du}{}^{\prime}f(u)\,\dfrac{\partial}{\partial x}f\left(\dfrac{y}{x}\right)$.

Section 4

1. (a) $\partial x/\partial a = 2a + b$; $\partial x/\partial b = a + 3b^2$; (b) $MP_a = 10$; $MP_b = 16$.
3. $\partial x/\partial L = 0.7575L^{-0.25}C^{0.25}$; $\partial x/\partial C = 0.2525C^{-0.75}L^{0.75}$.

Chapter 10

Section 2

1. $kz = \dfrac{a_1kx + b_1ky}{a_2kx + b_2ky} = \dfrac{a_1x + b_1y}{a_2x + b_2y}$;

(b) (1) $kz = akx + bky = k(ax + by)$;

(2) $kz = \sqrt{ak^2x^2 + 2hkxky + bk^2y^2} = k\sqrt{ax^2 + 2hxy + by^2}$;

(3) $kz = ak^a x^a k^{1-a} y^{1-a} = k(ax^a y^{1-a})$;

(4) $kz = \dfrac{ak^2x^2 + 2hkxky + bk^2y^2}{ckx + dky} = \dfrac{k(ax^2 + 2hxy + by^2)}{cx + dy}$;

(c) $k^2z = ak^2x^2 + 2hkxky + bk^2y^2 = k^2(ax^2 + 2hxy + by^2)$;

(d) $k^{a+b}z = ak^a x^a k^b y^b = k^{a+b}(ax^a y^b)$.

3. $k^3z = k^3(x^3 + 6x^2y - 3xy^2 + y^3)$.

5. $k^2z = k^2(3x^2 - 2y^2 - 6z^2 + 4u^2)$; $2^2z = 2^2(3x^2 - 2y^2 - 6z^2 + 4u^2)$.

Section 4

1. (a) $k^4z = k^4(3x^4 + 6y^4)$; (b) $x(\partial z/\partial x) + y(\partial z/\partial y) = 4z$.

3. (a) $k^2z = k^2(3a^2 + 4ab + b^2)$;

(b) $a(\partial z/\partial a) + b(\partial z/\partial b) = 2(3a^2 + 4ab + b^2)$; (c) 112; (d) 22;
(e) 110; (f) 2.

Chapter 11

Section 2

1. $-\frac{2}{3}$. **3.** $-7\frac{1}{3}$. **5.** ∞. **7.** 1. **9.** 1.

Section 4

1. $\dfrac{ED_a}{Ep_a} = \dfrac{-5p_a}{60 - 5p_a + 10p_b}$; $\dfrac{ED_a}{Ep_b} = \dfrac{+10p_b}{60 - 5p_a + 10p_b}$;

(b) $\dfrac{ED_a}{Ep_a} = \dfrac{-5}{6}$; $\dfrac{ED_a}{Ep_b} = \dfrac{5}{6}$.

3. $\dfrac{ED_a}{Ep_a} = \dfrac{-1.9p_a}{63.3 - 1.9p_a + 0.2p_b + 0.5p_c}$;

$\dfrac{ED_a}{Ep_b} = \dfrac{0.2p_b}{63.3 - 1.9p_a + 0.2p_b + 0.5p_c}$;

$\dfrac{ED_a}{Ep_c} = \dfrac{0.5p_c}{63.3 - 1.9p_a + 0.2p_b + 0.5p_c}$.

5. $Ex/Ep = -1.3$; $Ex/Eq = 0.4$; $Ey/Ep = 0.3$; $Ey/Eq = -0.1$.

Chapter 12

Section 2

1. (a) $e = \left(\dfrac{p}{R'-p}\right)$; (b) $p = R'\left(\dfrac{e}{e+1}\right)$.

3. (a) $R=2x-5x^2+3x^3$; (b) $e = \dfrac{2-5x+3x^2}{-5x+6x^2}$;

(c) $R'=2-10x+9x^2$.

5. $R'=p$. **7.** ∞. **9.** -3.33. **11.** $-9/5$.

Chapter 13

Section 2

1. $\frac{1}{4}x^4+C$. **3.** $x-x^2+C$. **5.** $-1/x+C$.

Section 5

1. At $x=4$, $z=32$. **3.** (a) $12-2x$; (b) 3. **5.** $\int MC = TVC$.

Section 7

1. x^3-4x^2+8x+4. **3.** x^3-4x^2+8x. **5.** $4/x$. **7.** $12-2x$.

9. 2. **11.** $\frac{4}{3}$. **13.** 2. **15.** 8. **17.** 8. **19.** 6. **21.** 16. **23.** -2.

Chapter 14

Section 2

7. (a) *AE*, *CF*, *AGKF*, *CKGE*; (b) *OS*.

Chapter 16

Section 2

3. 6. **13.** (a) b; (b) ab; (c) $P/(P-b)=OP/PT$ in basic elasticity formula; (d) -1; (f) $\frac{3}{4}b$.

Chapter 17

Section 2

1. (a) (1) $-\frac{3}{2}$; (2) $-\frac{5}{4}$; (3) $-\frac{6}{5}$; (4) -1; (5) $-\frac{11}{9}$.

3. 0. **5.** Arc elasticity of each curve $= -1$.

Section 6

5. 13/11.

7. When slope of AC is negative, $E_{tc} < 1$; when slope of $AC=0$, $E_{tc}=1$; when slope of AC is positive, $E_{tc} > 1$.

Chapter 19

Section 4

13. (a) R/OM; (b) 0; consumer must have at least OM of X plus some of Y or at least ON of Y plus some of X; (d) no; (e) no.

Chapter 21

Section 4

1. (c) 6 of X and 8 of Y to produce 100; 9 of X and 12 of Y to produce 150; (e) $\frac{4}{3}$; (g) $\frac{3}{4}$.

3. (d) Process B; 80 of X_1 and 40 of X_2 to produce 400; (e) A or B; (f) A.

Index

O

P

Q

R

S

T